MEASUREMENTS AND CORRELATION FUNCTIONS

MEASUREMENTS AND CORRELATION FUNCTIONS

PAUL C. MARTIN
Professor of Physics, Harvard University
Cambridge, Massachusetts

GORDON AND BREACH SCIENCE PUBLISHERS
NEW YORK · LONDON · PARIS

Copyright © 1968 by GORDON AND BREACH, Science Publishers, Inc.
150 Fifth Avenue, New York, N.Y. 10011

Library of Congress Catalog Card Number 68-18964

Editorial office for Great Britain:
Gordon and Breach, Science Publishers Ltd.
12 Bloomsbury Way
London W.C.1

Editorial office for France:
Gordon & Breach
7-9 rue Emile Dubois
Paris 14e

Distributed in Canada by:
The Ryerson Press
299 Queen Street West
Toronto 2B, Ontario

Printed in the United States of America

INTRODUCTION

Whether a system has a few or a very large number of dynamical degrees of freedom, it is difficult, and would probably not be very rewarding, to measure complicated correlations between many physical observables. In particular, the precise wave function or density matrix of a large system (a function of many variables) or the corresponding classical phase space density is of no concern. Although these statements are obvious to anyone who really measures or calculates physical properties of a large interacting system, the dynamics of interacting systems were so poorly understood for so many years that no objections were raised to their use. The wave function and density matrix were discussed as if these detailed and infinitely complicated functions of infinitely many variables really mattered.

Indeed, when field correlation functions were introduced in the fifties to describe the properties of interacting systems, they were viewed as formal devices and treated with some suspicion and derision. Reasons for this uneasiness were not difficult to find. The functions were introduced along with calculational techniques that were both sophisticated and complicated; the nature of the experimental information they contained was either not fully explained or not fully understood; and this experimental information was often not accessible. By now, however, correlation functions, the natural and indispensable tools of electrical engineers and particle physicists, have been almost universally adopted for discussing problems in solid state and statistical physics.

I have always intended to write a treatise on many particle physics which began by illustrating the importance, convenience, and physical significance of the correlation function with the aid of simple examples. A description of the properties of different physical systems in terms of correlations of physical observables ought to be as natural to the student or experimenter as it is to the theoretical physicist. I hoped to go on to discuss classical and quantum interacting fluids, plasmas, solids, magnetic systems, superfluids and superconductors in terms of correlation functions. In each system I would characterize the order, the elementary excitations, and the collective modes in terms of simple physically measurable correlation functions. Unfortunately I have never passed far beyond the introduction of that book. Indeed, when I look

v

at this version of the introduction, lectures I presented at Harvard and Les Houches, I can see many ways of improving and simplifying it.

If I were to write a treatise, I would also have to cite the parallel work of many authors in many schools — Kubo, Landau, Luttinger, Matsubara, Prigogine, the Saclay group, van Hove, Zwanzig, and many others. Since this is just a set of lecture notes attempting to explain a physical approach to experimenters, I hope they and many other authors and friends will forgive me for omitting references to their work.

CONTENTS

TO MY STUDENTS

Without them I would have had more time to write but less to write about. Their questions and ideas have contributed immensely to statistical physics and to my understanding of it.

A. Illustrative Example

In these lectures we will try to indicate the importance and inevitability of the correlation function description of physical measurements. As an introduction to their properties and significance let us speak about a familiar simple example: a single specified degree of freedom, say a one dimensional oscillator, interacting with its surroundings. We might for example be concerned literally with an oscillator, or a heavy particle (an oscillator with spring constant zero) being bombarded by the molecules of a gas or liquid in which it is placed, or we might be concerned with some idealization of a charge dipole in the presence of black body radiation. The effect of the medium will be to add an internal force $F(t)$ to the restoring force of the spring so that, associated with Fig. 1, we have Newton's equation

$$m\ddot{x}(t) + m\omega_0^2 x(t) = F^{\text{int}}(t). \tag{1}$$

On the average, or for a collection of such oscillators, the expectation value of each term in equation (1) will vanish. However, if at some particular time the oscillator has a known infinitesimal displacement from equilibrium, then the average behavior at later times will exhibit decay, and perhaps oscillation. It is this conditional average and quantities related to it which we ordinarily discuss and measure.

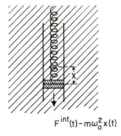

$$F^{\text{int}}(t) - m\omega_0^2 x(t)$$

Fig. 1. Oscillator fluctuating in equilibrium. $x(t)$ is the instantaneous displacement of the center of mass from its equilibrium value, $F^{\text{int}}(t)$ the instantaneous force due to the medium, and $-m\omega_0^2 x(t)$ the instantaneous force of the spring.

One natural way of describing the behavior of the system is to employ the function which relates the infinitesimal displacement of the oscillator to the infinitesimal external force by which we would physically displace it. If we call the external force $F^{\text{ext}}(t)$, the resultant average non-equilibrium

1

displacement $\langle x(t) \rangle_{\text{n.e.}}$, and the function which linearly relates them $\tilde{\chi}(tt')$, we may write

$$\langle x(t) \rangle_{\text{n.e.}} = \int_{-\infty}^{\infty} \tilde{\chi}(tt') \, F^{\text{ext}}(t') \, dt'. \tag{2}$$

In the presence of the additional applied force, equation (1) becomes

$$m\ddot{x}(t) + m\omega_0^2 x(t) = F^{\text{int}}(t) + F^{\text{ext}}(t) \tag{3}$$

so that on the average

$$m\langle \ddot{x}(t) \rangle_{\text{n.e.}} + m\omega_0^2 \langle x(t) \rangle_{\text{n.e.}} - \langle F^{\text{int}}(t) \rangle_{\text{n.e.}} = F^{\text{ext}}(t). \tag{4}$$

Now, however $\langle F^{\text{int}}(t) \rangle_{\text{n.e.}}$ does not vanish. Indeed, phenomenological laws which describe the physical properties of the medium are ordinarily incorporated by telling how $\langle F^{\text{int}}(t) \rangle_{\text{n.e.}}$ depends on the displacement. Once we have expressed $\langle F^{\text{int}}(t) \rangle_{\text{n.e.}}$ in terms of $\langle x(t') \rangle_{\text{n.e.}}$, we can solve for $\langle x(t) \rangle_{\text{n.e.}}$ in terms of $F^{\text{ext}}(t)$, so determining the $\tilde{\chi}(tt')$ we measure. For example, for many purposes it is sufficient to say that the medium is described by a friction constant and that

$$\langle F^{\text{int}}(t) \rangle_{\text{n.e.}} \approx -m\gamma \langle \dot{x}(t) \rangle_{\text{n.e.}}. \tag{5}$$

Let us temporarily illustrate our general discussion using this simple law although eventually we shall find that it is inadequate. Let us also keep in mind the electromagnetic example where the simple phenomenological law for radiation damping is

$$\langle F(t) \rangle_{\text{n.e.}} \approx m\left(\frac{e^2}{6\pi mc^3} \right) \langle \dddot{x}(t) \rangle_{\text{n.e.}}. \tag{5'}$$

The phenomenological equation for $\langle F^{\text{int}}(t) \rangle_{\text{n.e.}}$, implies corresponding equations for the quantities $\langle x(t) \rangle_{\text{n.e.}}$

$$m\langle \ddot{x}(t) \rangle_{\text{n.e.}} + m\omega_0^2 \langle x(t) \rangle_{\text{n.e.}} + m\gamma \langle \dot{x}(t) \rangle_{\text{n.e.}} = F^{\text{ext}}(t) \tag{6}$$

and (in view of Eq. (2)), for $\tilde{\chi}(tt')$,

$$m\frac{d^2}{dt^2} \tilde{\chi}(tt') + m\omega_0^2 \tilde{\chi}(tt') + m\gamma \frac{d}{dt} \tilde{\chi}(tt') = \delta(t - t'). \tag{7}$$

That is to say in this example $\tilde{\chi}(tt')$ is the Green's function for the ordinary differential equation describing the non-equilibrium behavior. Specifically it gives the response to a unit impulsive external force at time t'. Both $\langle x(t) \rangle_{\text{n.e.}}$ for a general $F^{\text{ext}}(t)$ and $\tilde{\chi}(tt') = \tilde{\chi}(t - t')$ vanish until an external force is applied. $\tilde{\chi}(t - t')$ is therefore called the retarded response function, or Green's function. The function that describes the solution which vanishes *after* the impulse is applied, $\chi(t - t')$, is called the advanced response function, or Green's function.

Both in the specific model, and more generally, it is often convenient to discuss, and practical to measure the Fourier transform of $\tilde{\chi}(t - t')$

$$\tilde{\chi}(t - t') = \int_{-\infty}^{\infty} \frac{d\omega}{2\pi} e^{-i\omega(t-t')} \chi(\omega) \tag{8}$$

$$\chi(\omega) = \int_{-\infty}^{\infty} d(t - t') e^{i\omega(t-t')} \tilde{\chi}(t - t') = \int_{0}^{\infty} d(t - t') e^{i\omega(t-t')} \tilde{\chi}(t - t') \tag{9}$$

where the last equality reflects the retarded or causal nature χ. It is also convenient to define $\chi(z)$, a function of a complex variable z, for z in the upper half complex plane. Clearly the last integral provides such a definition since it is well defined and exponentially decreasing when $\omega \to z = \omega + i\varepsilon$ ($\varepsilon > 0$). The function of a complex variable z, $\chi(z) \to \chi(\omega)$ as $\varepsilon \to 0$. $\chi(z)$ so defined in the upper half z plane, is clearly analytic and bounded there. By inspection of Eq. (7) we see that $\chi(z)$ satisfies

$$m(-z^2 + \omega_0^2 - i\gamma z) \chi(z) = 1. \tag{10}$$

We may immediately invert this equation if we wish to, and write

$$\tilde{\chi}(t - t') = \int_{-\infty}^{\infty} \frac{d\omega}{2\pi} \frac{e^{-i\omega(t-t')}}{(-\omega^2 + \omega_0^2 - i\gamma\omega) m} \tag{11}$$

where we integrate along the real axis (or any path from $-\infty$ to ∞ in the upper half plane).

Explicitly we have for the response to an impulse the function illustrated in Fig. 2

$\tilde{\chi}(t - t')$

$= \eta(t - t') \, m^{-1} \exp\left[-\tfrac{1}{2}\gamma(t - t')\right] \sin\left\{[\omega_0^2 - \tfrac{1}{4}\gamma^2]^{\frac{1}{2}} (t - t')\right\} [\omega_0^2 - \tfrac{1}{4}\gamma^2]^{-\frac{1}{2}}$

where $\eta(t - t') = 1$ when $t > t'$ and 0 when $t < t'$.

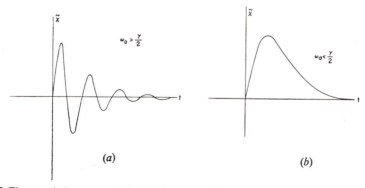

(a)

(b)

Fig. 2. The retarded response, $\tilde{\chi}(t)$. (a) When $\omega_0 > \gamma/2$ (underdamped), the mass oscillates; (b) when $\omega_0 < \gamma/2$ (overdamped), it decays without oscillation.

For real ω, the response function, $\chi(\omega)$, is usually divided into two parts, a dissipative response and a reactive response. In our example, and more generally when the system is time-reversal invariant, these are described respectively by the imaginary and real parts of $\chi(\omega)$, and denoted as $\chi''(\omega)$ and $\chi'(\omega)$.

In the phenomenological model

$$\chi''(\omega) = \frac{\omega\gamma}{m[(\omega^2 - \omega_0^2)^2 + (\gamma\omega)^2]} \,. \tag{12}$$

The Fourier transform of $\chi''(\omega)$ is the imaginary odd function of time illustrated in Fig. 3

$$\tilde{\chi}''(t - t') = \int \frac{d\omega}{2\pi} e^{-i\omega(t-t')} \chi''(\omega) = \frac{1}{2i} [\tilde{\chi}(t - t') - \tilde{\chi}(t' - t)]$$

$$= (2im)^{-1} \exp\left[-\tfrac{1}{2}\gamma |t - t'|\right] \sin\left\{\omega_0^2 - \tfrac{1}{4}\gamma^2\right]^{1/2} (t - t')\right\} [\omega_0^2 - \tfrac{1}{4}\gamma^2]^{-1/2}$$

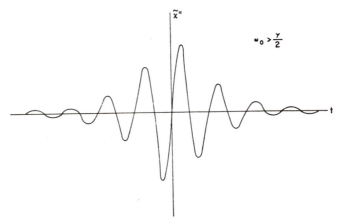

Fig. 3. The absorptive response, $\chi''(t)$. The absorptive response is purely imaginary; plotted is $2i\tilde{\chi}''(t)$ for the underdamped situation in Fig. 2a.

In general, $\tilde{\chi}''$ is defined as the difference between advanced and retarded functions,

$$\tilde{\chi}''(t - t') = \tfrac{1}{2}i [\tilde{\chi}(t - t') - \tilde{\chi}(t = t')]$$

When the system is not time reversal invariant, $\tilde{\chi}(t - t') \neq -\tilde{\chi}(t' - t)$, and $\tilde{\chi}''$ can have an even part in $t = t'$. In that case, $\chi''(\omega)$ will also have an imaginary part.

Likewise in the model, we have

$$\chi'(\omega) = \frac{\omega_0^2 - \omega^2}{m[(\omega^2 - \omega_0^2)^2 + (\gamma\omega)^2]} \tag{13}$$

whose Fourier transform is the real even function of time illustrated in Fig. 4

$$\tilde{\chi}'(t - t') = \frac{1}{2} [\tilde{\chi}(t - t') + \tilde{\chi}(t' - t)]$$

$$= (2m)^{-1} \exp\left[-\tfrac{1}{2}\gamma|t - t'|\right] \sin\left\{[\omega_0^2 - \tfrac{1}{4}\gamma^2]^{1/2} |t - t'|\right\} [\omega_0^2 - \tfrac{1}{4}\gamma^2]^{-1/2}.$$

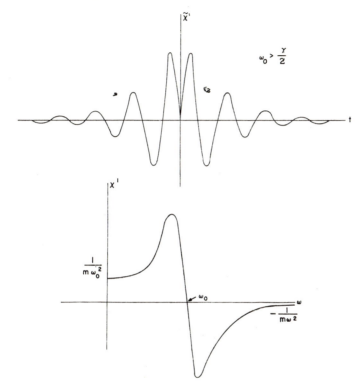

Fig. 4. The dispersive response $\tilde{\chi}'(t)$ and its transform $\chi(\omega)$ for the underdamped oscillator. The former is the even function corresponding to $\tilde{\chi}(t)$. The latter is also even and has the characteristic change in sign at ω_0.

To see that $\chi''(\omega)$ represents the dissipation let us calculate the work done by our infinitesimal external force

$$-\frac{dW}{dt} = F^{\text{ext}}(t) \langle v(t) \rangle_{\text{n.e.}} = F^{\text{ext}}(t) \langle \dot{x}(t) \rangle_{\text{n.e.}}$$

$$= \int_{-\infty}^{\infty} \frac{d\bar{\omega}}{2\pi} e^{-i\bar{\omega}t} F^{\text{ext}}(t) (-i\bar{\omega}) \chi(\bar{\omega}) \int_{-\infty}^{\infty} dt' e^{i\bar{\omega}t'} F^{\text{ext}}(t'). \tag{14}$$

For a monochromatic forcing term $F^{ext}(t) = F^{ext} \cos(\omega t + \theta) = \text{Re}(\mathscr{F}^{ext} e^{i\omega t})$ the time average value over a cycle is

$$-\frac{\overline{dW}}{dt} = \frac{1}{2}\omega \, \text{Im} \, \chi(\omega) \mathscr{F}^{ext*} \mathscr{F}^{ext} = \frac{1}{2}\omega\chi''(\omega)\mathscr{F}^{ext*}\mathscr{F}^{ext}$$

$$= \omega\chi''(\omega)\,\overline{F^2(t)}. \tag{15}$$

In deriving these expressions, we have used the oddness of $\chi''(\omega)$, which is true in our model and much more generally. Note, incidentally, that the dissipation in the electromagnetic illustration

$$\omega\chi''(\omega) = \frac{e^2\omega^4}{6\pi mc^3}\left[(\omega^2 - \omega_0^2)^2 + \left(\frac{e^2\omega^3}{6\pi mc^3}\right)^2\right]^{-1}$$

behaves as ω^4 in accordance with our expectation from the Rayleigh scattering law.

As $\omega \to 0$, $\chi'(0) = \chi(0) = 1/m\omega_0^2$, which indicates that the constant displacement resulting from a time independent force is $x = F^{ext}/m\omega_0^2$ and produces no dissipation. As $\omega \to \infty$, $\chi'(\omega) = \chi(\omega) = -1/m\omega^2$; the oscillator responds independent of its surroundings, and no dissipation occurs. In this case the oscillator is 180^0 out of phase with the force. The two limits are perhaps most familiar as limits of the electric polarisability divided by ne^2 in a dielectric.

Because the response is causal, or equivalently, because $\chi(z)$ which we have defined in the upper half plane is analytic there, the real and imaginary parts of $\chi(\omega)$ are related by the Kramer's-Kronig relations

$$\chi'(\omega) = P\int_{-\infty}^{\infty} \frac{d\omega'}{\pi}\frac{\chi''(\omega')}{\omega' - \omega}$$

$$\chi''(\omega) = -P\int_{-\infty}^{\infty} \frac{d\omega'}{\pi}\frac{\chi'(\omega')}{\omega' - \omega}. \tag{16}$$

P means principal value integral, that is, an integral symmetrical about the singularity.

Rather than proving these familiar formulas in the most direct manner let us derive them from an equivalent expression which is also valid for complex z. In particular let us express $\chi(z)$ for complex z in terms of $\chi''(\omega)$. When z is in the upper half plane, we may use Cauchy's theorem on the contour shown in Fig. 5

$$\chi(z) = \int_{-\infty}^{\infty}\frac{d\omega}{2\pi i}\frac{\chi(\omega)}{\omega - z} = \int_{-\infty}^{\infty}\frac{d\omega}{2\pi i}\chi(\omega)\left[\frac{1}{\omega - z} \pm \frac{1}{\omega - z^*}\right]. \tag{17}$$

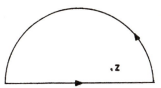

Fig. 5. Contour for Cauchy's Theorem. Since $\chi''(\omega)$ vanishes rapidly as $\omega \to \infty$, the semicircle at infinity does not contribute.

We have added or subtracted the last term which is identically zero since $\chi(z')/(z' - z^*)$ is analytic in the upper half z' plane. Hence

$$\chi(z) = \int_{-\infty}^{\infty} \frac{d\omega}{\pi i} \chi(\omega) \, \text{Re} \, \frac{1}{\omega - z} = \int_{-\infty}^{\infty} \frac{d\omega}{\pi} \chi(\omega) \, \text{Im} \, \frac{1}{\omega - z}. \tag{18}$$

If we add the real part of $\chi(z)$ determined from the first identity and the imaginary part of $\chi(z)$ from the second we find

$$\chi(z) = \int_{-\infty}^{\infty} \frac{d\omega}{\pi} \, \text{Im} \, \chi(\omega) \left[\text{Re} \frac{1}{\omega - z} + i \, \text{Im} \frac{1}{\omega - z} \right], \tag{19}$$

so that for z in the upper half plane

$$\chi(z) = \int_{-\infty}^{\infty} \frac{d\omega}{\pi} \frac{\chi''(\omega)}{\omega - z}. \tag{20}$$

We shall take this expression to define $\chi(z)$ throughout the complex z plane. Clearly, if $z = \omega + i\varepsilon$ and $\varepsilon \to 0$, we recover the Kramers-Kronig relation since

$$\chi(\omega) = \lim_{\varepsilon \to 0} \int \frac{d\omega'}{\pi} \frac{\chi''(\omega')}{\omega' - (\omega + i\varepsilon)} = P \int \frac{d\omega'}{\pi} \frac{\chi''(\omega')}{\omega' - \omega} + i \chi''(\omega) \tag{21}$$

because of the identity

$$\frac{1}{x \mp i\varepsilon} = \frac{x \pm i\varepsilon}{x^2 + \varepsilon^2} = P \frac{1}{x} \pm \pi \, i\delta(x).$$

Note that the function $\chi(z)$ is analytic except on the real axis where there is a branch line. In the phenomenological example we are using as an illustration, $\chi(z)$ is given by (10) in the upper half plane but by

$$\chi(z) = [m(-z^2 + \omega_0^2 + i\gamma z)]^{-1} \tag{22}$$

in the lower half plane. There is no pole in either half plane but a branch line along the real axis across which there is a discontinuity $2\chi''(\omega)$. In the phenomenological example, it is possible to continue the function $\chi(z)$ onto another sheet. The continuation of the function from the upper half plane has a pole as does the continuation from the lower half plane. The existence of such continuations depends on there being a branch line in $\chi(z)$. Such a branch line may either be the result of considering infinite systems or of what is sometimes called "time smoothing." Were we speaking accurately and in a microscopic fashion about the quantity $\chi(z)$ we would discover that on a gross scale we could describe it in a similar manner. However, if we resolved the function very carefully mathematically we would find not a branch line but a sequence of physically unresolvable discrete closely spaced poles. Correspondingly $\chi''(\omega)$ would be a sequence of discrete closely spaced δ-functions. See Figs. 6 and 7.

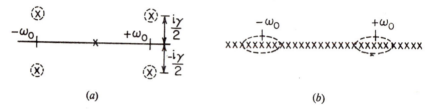

(a)   (b)

Fig. 6. (*a*) Structure of smoothed function $\chi(z)$ in complex plane. The function has no poles but its analytic continuation has poles in both the upper and lower half plane on a second sheet.

(*b*) Structure of unsmoothed $\chi(z)$ in complex plane. There are very closely spaced poles corresponding to all possible energy differences of the large system. On the average, the residues of the poles near $\pm\omega_0$ are larger than elsewhere. The poles merge into the line singularity in (a) upon smoothing.

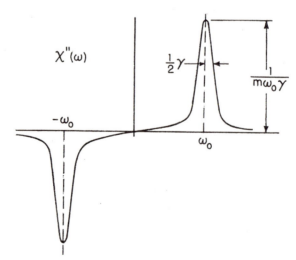

Fig. 7a. Discontinuity of smoothed function $\chi(z)$ is equal to $2\chi''(\omega)$.

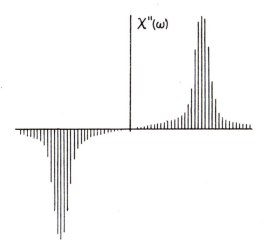

Fig. 7b. Schematic indication of discontinuity in unsmoothed $\chi''(\omega)$, i.e., of δ-functions at the closely spaced poles, with infinitesimal residues which are on the average largest when $\omega \sim \omega_0$.

As a result, at times long compared to the frequency differences between these discrete poles (whose distance goes rapidly to zero with the number of degrees of freedom or volume) we would have recurrent behavior. There would be no continuation. Thus the symmetrical irreversibility (the absence of recurrence) is introduced when we represent the discrete oscillatory behavior resulting from the poles by the equivalent behavior for times short compared to the inverse of the infinitesimal frequency differences of the continuous distribution of poles or branch line (i.e. when we replace a sum by an integral). This aspect, which is valid as the number of degrees of freedom becomes infinite, does not remove the symmetry between t and $-t$. The equations and functions like $\tilde{\chi}'(t - t')$ and $\tilde{\chi}''(t - t')$ are still perfectly symmetrical.

The asymmetric aspect, the "arrow of time," has here to do with the physically realizable boundary conditions on the differential equations – – the fact that we perform experiments in which we take a system from equilibrium in a finite time interval and it relaxes over essentially infinite times. It is this aspect of the experimentally appropriate boundary conditions which leads us to use the retarded function $\tilde{\chi}(t - t') = 2i\eta(t - t')\,\tilde{\chi}''(t - t')$. If we could perform an experiment in which the final rather than the initial conditions were specified, we would use a different solution.

These are the only statements I shall make about such "philosophical" questions, but if I didn't make them, I am sure I would be called to task later.

Let us note a more practical aspect of our formula for $\chi(z)$, namely, the expression which results from expansion at short times or high frequencies

$$\chi(z) \sim -\frac{1}{z} \sum_{n=0}^{\infty} \int_{-\infty}^{\infty} \frac{d\omega}{\pi} \chi''(\omega) \left(\frac{\omega}{z}\right)^n \qquad (23)$$

and which is valid for large enough z provided

$$\int \frac{d\omega}{\pi} \chi''(\omega) \omega^n \qquad (24)$$

converges. Comparing coefficients for large z, we have for our phenomenological models

$$\chi(z) \sim -\frac{1}{mz^2} = -\frac{1}{z^2} \int_{-\infty}^{\infty} \frac{d\omega}{\pi} \chi''(\omega) \omega. \qquad (25)$$

This is a well-known sum-rule, which is generally valid, i.e.,

$$\int_{-\infty}^{\infty} \frac{d\omega}{\pi} \chi''(\omega) \omega = \frac{1}{m}. \qquad (26)$$

For our example, we have

$$\int \frac{d\omega}{\pi} \frac{\omega^2 \gamma}{[(\omega^2 - \omega_0^2)^2 + (\gamma\omega)^2] m} = \frac{1}{m}.$$

If we take our phenomenological expressions seriously, however, we see that they give a divergent result for

$$\int \frac{d\omega}{\pi} \chi''(\omega) \omega^3.$$

We may begin to understand why this is a failing of the phenomenological models by recalling the derivation of the radiation damping expression. In that derivation one assumes that $\omega e^2/mc^3 \ll 1$. For times short (frequencies high) compared to the time it takes a light signal to cross a classical electron, one must consider the actual radiation processes more carefully. When one does, the damping no longer increases like the cube of the frequency. In fact, for any system, the relation between the internal force and the displacement must vanish at high frequencies. It takes time for the medium to respond to the fact we have forced the oscillator. Not only does the restoring force vanish at high frequencies; it vanishes faster than any power of the frequency so that in a microscopic theory, and in a physical measurement of $\chi''(\omega)$, we have a free oscillator as $\omega \to \infty$;

$$\int_{-\infty}^{\infty} \frac{d\omega}{\pi} \chi''(\omega) \omega^n$$

is finite for all n. ($\chi''(\omega) \to \pi\delta(\omega^2 - \omega_0^2) \omega/|\omega|$ plus small corrections.)

We can give a quick proof of these claims about the existence of all moments[2] by invoking a famous and useful theorem, the Nyquist theorem (or fluctuation-dissipation theorem) which we shall subsequently prove. According to this theorem the dissipation that results when an external field is applied to the system is simply related to the fluctuations in thermodynamic equilibrium. Thus our discussion of $\chi''(\omega)$ or $\chi(\omega)$ is literally as indicated in the first paragraph, a calculation of the behavior of the oscillator when no external forces are applied.

Specifically, the Nyquist theorem says that

$$\langle x(t)\, x(t')\rangle_{\text{eq.}} - \langle x(t)\rangle_{\text{eq.}} \langle x(t')\rangle_{\text{eq.}} = (\langle x(t)\, x(t')\rangle_{\text{eq.}})$$

$$= \int \frac{d\omega}{2\pi}\, e^{-i\omega(t-t')}\, 2\varepsilon(\omega)\, \chi''(\omega)/\omega \tag{27}$$

where $\varepsilon(\omega)$ is the mean energy of an oscillator with natural frequency ω at temperature $kT = \beta^{-1}$, that is,

$$\varepsilon(\omega) \equiv \hbar\omega \left[\frac{1}{2} + \frac{1}{e^{\beta\hbar\omega} - 1}\right] \underset{\text{class.}}{\to} \frac{1}{\beta}. \tag{28}$$

Thus the statement we previously indicated was a general sum rule

$$\int \frac{d\omega}{\pi}\, \chi''(\omega)\, \omega = \frac{1}{m} \tag{29}$$

is at least classically, just the statement

$$\frac{1}{2}\, m\langle \dot{x}^2(t)\rangle = \int \frac{d\omega}{2\pi}\, \frac{m}{\beta}\, \omega\chi''(\omega) = \frac{1}{2}\, kT \tag{30}$$

i.e., the statement that the mean kinetic energy of the oscillator is $\frac{1}{2}kT$.

But by exactly the same kind of argument, we have

$$\langle \ddot{x}^2(t)\rangle = \int \frac{d\omega}{\pi}\, \frac{\omega^3}{\beta}\, \chi''(\omega) \tag{31}$$

$$\left\langle \left(\frac{d^n x(t)}{dt^n}\right)^2\right\rangle = \int \frac{d\omega}{\pi}\, \frac{\omega^{2n-1}}{\beta}\, \chi''(\omega) \tag{32}$$

and the thermodynamic average of the squares of the higher derivatives of the position are all finite. One can, for example, calculate $\langle \ddot{x}^2(t)\rangle$ directly for an oscillator interacting with particles by a potential $\sum_\alpha v(x(t) - x_\alpha(t))$, deducing

$$\langle \ddot{x}^2(t)\rangle \equiv \frac{kT}{m}\, \omega_\infty^2 \equiv \frac{kT}{m}\, \langle \omega_{vv}^2\rangle$$

$$\omega_\infty^2 = \frac{n}{3m} \int d^3 r g(r)\, \nabla^2 v(r) + \omega_0^2 \tag{33}$$

where $g(r)$ is the equilibrium correlation function between the medium and oscillator.

It is possible to incorporate this "sum rule" by modifying our simplest phenomenological model. In particular we may introduce a description like the ones employed by Drude and Maxwell at the turn of the century. We suppose that the induced internal force satisfies the phenomenological law

$$\frac{1}{(\omega_\infty^2 - \omega_0^2)} \frac{\partial \delta \langle F^{\text{int}} \rangle_{\text{n.e.}}}{\partial t} + \frac{\delta \langle F^{\text{int}} \rangle_{\text{n.e.}}}{\gamma} = -\frac{m \partial \delta \langle x \rangle_{\text{n.e.}}}{\partial t} \tag{34}$$

which interpolates between the high frequency reactive behavior characterized by (33), and the low frequency phenomenological damping when $\omega << \tau \equiv (\omega_\infty^2 - \omega_0^2)/\gamma$, or

$$\delta \langle F^{\text{int}} \rangle_{\text{n.e.}} = \frac{m i \omega \tau}{1 - i \omega \tau} (\omega_\infty^2 - \omega_0^2) \, \delta \langle x \rangle_{\text{n.e.}} . \tag{35}$$

This phenomenological law, which of course is still not really adequate since it predicts

$$\int \omega^5 \chi''(\omega) \frac{d\omega}{\pi} = \infty,$$

gives a familiar kind[3] of expression for $\chi(z)$, namely, for $\text{Im } z > 0$,

$$\chi^{-1}(z) = -m \left[z^2 - \omega_0^2 + \frac{(\omega_\infty^2 - \omega_0^2) \tau z i}{1 - i z \tau} \right]; \quad \gamma = (\omega_\infty^2 - \omega_0^2) \tau . \tag{36}$$

Instead of introducing successively more satisfactory phenomenological descriptions which will always be somewhat ad hoc, let us write the almost tautological equation (which we shall also discuss more formally at a later stage)

$$\chi^{-1}(z) \equiv -m[z^2 - \omega_0^2 + i z \gamma(z)]. \tag{37}$$

In this equation we have replaced the unknown response function $\chi(z)$ by an equally unknown function $\gamma(z)$

$$\gamma(z) = \int \frac{d\omega'}{\pi i} \frac{\gamma'(\omega')}{\omega' - z} \to \gamma'(\omega) + i\gamma''(\omega) \quad \text{as } z = \omega + i\varepsilon \to \omega \tag{38}$$

$$\gamma''(\omega) = -P \int \frac{d\omega'}{\pi} \frac{\gamma'(\omega')}{\omega' - \omega} \tag{39}$$

in which $\gamma(z)$ is associated with the phenomenological law. For an oscillator of the type considered, symmetry properties require that $\gamma'(\omega)$ is real, even and positive so the oscillator is described by

$$\chi^{-1}(z) = -m \left[z^2 - \omega_0^2 + z^2 \int \frac{d\omega}{\pi} \frac{\gamma'(\omega)}{\omega^2 - z^2} \right] \tag{40}$$

or

$$\chi''(\omega) = \frac{1}{m} \frac{\omega\gamma'(\omega)}{(\omega^2 - \omega_0^2 - \omega\gamma''(\omega))^2 + (\omega\gamma'(\omega))^2} \qquad (41)$$

These expressions in terms of γ rigorously describe the properties of an oscillator coupled to its surroundings in any time-reversal invariant system. The great variety of possible behaviors manifests itself in the diverse possibilities for $\gamma(z)$. Depending on the coupling there may be one or many "renormalized" natural frequencies, $\bar{\omega}$, of the oscillator, that is, solutions to the equation

$$\bar{\omega}^2 - \omega_0^2 - \bar{\omega}\gamma''(\bar{\omega}) = 0. \qquad (42)$$

These solutions will be of interest if the quantity $\omega\gamma'(\omega)$ is small and slowly varying near $\bar{\omega}$ since, in the neighborhood of $\bar{\omega}$, they correspond to resonances.[4] They will be true normal modes of the oscillator if $\gamma'(\bar{\omega}) = 0$.

In the neighborhood of a relatively well defined mode or *resonant frequency*, $\bar{\omega}$, we may write

$$\chi''(\omega) \cong \frac{1}{m} \frac{Z(\bar{\omega})\,\bar{\omega}\bar{\gamma}'(\bar{\omega})}{(\omega^2 - \bar{\omega}^2)^2 + (\bar{\omega}\bar{\gamma}'(\bar{\omega}))^2}$$

$$\omega\chi''(\omega) \cong \frac{1}{2m} \frac{Z(\bar{\omega})\tfrac{1}{2}\bar{\gamma}'(\bar{\omega})}{(\omega - \bar{\omega})^2 + (\tfrac{1}{2}\bar{\gamma}'(\bar{\omega}))^2} \qquad (43)$$

where we have introduced

$$\bar{\gamma}'(\bar{\omega}) = Z(\bar{\omega})\,\gamma'(\bar{\omega}) \qquad (44)$$

and

$$Z^{-1}(\bar{\omega}) \cong 1 - \frac{\partial}{\partial\bar{\omega}^2}\,\bar{\omega}\gamma''(\bar{\omega}). \qquad (45)$$

The quantity $\tfrac{1}{2}\bar{\gamma}'(\bar{\omega})$ is the half-width at half-height of the resonance and describes the *rate* at which the ocillator amplitude decays. The energy of the oscillator, quadratic in the amplitude, decays at the rate $\bar{\gamma}'(\bar{\omega})$ in the neighborhood of the resonance. The quantity, $Z(\bar{\omega})$, the renormalization constant, represents the *strength* or fraction of the oscillator motion which participates in the approximate normal mode, $\bar{\omega}$, of the coupled system; or more precisely, (when $\bar{\gamma}'$ is small but not zero) in the many normal modes centered about $\bar{\omega}$. Because γ' is positive, $d\bar{\omega}\gamma''(\bar{\omega})/d\bar{\omega}^2 < 0$, so that the fraction, $Z(\bar{\omega})$, in any normal mode is less than unity. Describing each resonance, therefore, there are three parameters; resonant frequency, lifetime, and strength. For the uncoupled oscillator, $\bar{\omega} = \pm\omega_0$, $\bar{\gamma}'(\bar{\omega}) = 0$, and $Z(\bar{\omega}) = 1$. If the oscillator were coupled to a single other oscillator with frequency $\tilde{\omega}_0$ we would have[5]

$$\gamma'(\omega) = \pi\lambda\,\delta(\tilde{\omega}_0^2 - \omega^2).$$

There would then be two normal modes; two roots $\bar{\omega}_1^2$ and $\bar{\omega}_2^2$. Each would have infinite lifetime ($\bar{\gamma}'(\bar{\omega}_i) = 0$) and $Z(\bar{\omega}_1) + Z(\bar{\omega}_2) = 1$. See Fig. 8.

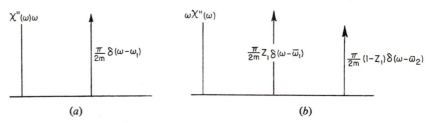

Fig. 8(a). The absorption $\omega\chi''(\omega)$ for an uncoupled oscillator. (b) The absorption $\omega\chi''(\omega)$ for an oscillator coupled to another oscillator, the normal frequencies of the pair being $\bar{\omega}_1$ and $\bar{\omega}_2$. The quantity Z_1 represents the fraction of the first oscillator displacement in the first (normalized) normal mode.

An example of the way a coupled oscillator exhibits both normal modes is provided in a recent experiment by one of the participants in these lectures, Dr. Wright. In an experiment in which Raman scattering by a longitudinal optical phonon was observed[6] he found that by altering the carrier concentration in GaAs he could alter the plasma frequency, ω_p, and the coupling of the plasma mode to the longitudinal optical mode, ω_l. The resultant $\chi(\omega)$ is schematically given by

$$\chi^{-1}(z) \propto \left[z^2 - \omega_l^2 + \frac{c\omega_p^2}{\omega_p^2 - z^2} \right]. \tag{46}$$

The variation of the two resultant peaks with ω_p^2 is shown in Fig. 9.

The weak coupling to infinitely many modes, by contrast, will frequently lead to a reduction in $Z(\bar{\omega})$ from unity at a single resonance, without the appearance of any additional resonance. These properties are depicted in Fig. 10.

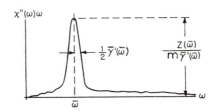

Fig. 10. The absorption in an oscillator coupled to many degrees of freedom, as in Fig. 7. The significance of the renormalized frequency $\bar{\omega}$, the half width $\frac{1}{2}\bar{\gamma}'(\omega)$, and the strength $Z(\bar{\omega})$, are shown.

The oscillator strength, originally lodged in the discrete mode is now shared among the infinitely many modes of the coupled system. Part of it, a fraction defined by Z, is shared in a fashion described by a Lorentzian, over nearby modes. The remainder is divided in a model dependent fashion.

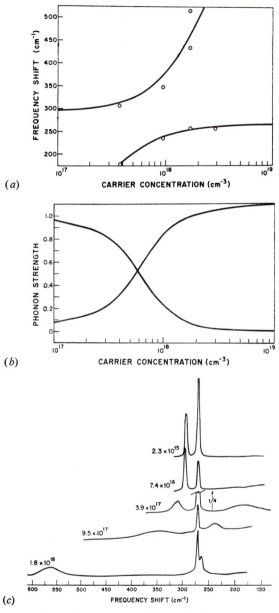

(a)

(b)

(c)

Fig. 9. Experimental illustration of coupled modes. Because the longitudinal optical mode in GaAs is coupled to the plasma mode according to Eq. (46), variation of the carrier concentration, n, and hence the plasma frequency $\omega_p^2 = [11.4 \times 10^{-14} n]$ cm^{-2} alters (a) the frequencies of the coupled modes $\overline{\omega}_1$ and $\overline{\omega}_2$, and (b) the strength, Z_1, in each mode. The other constants in Eq. (46) are $c = [1.29 \times 10^5]$ cm^{-2} and $\omega_l = 291$ cm^{-1}. In (c) some typical tracings are shown.

The simple model we have been discussing is not as experimentally accessible as some more complicated examples. The three dimensional analog, however, is exemplified by the motion of tagged particle (the oscillator) in a fluid (say a noble liquid) and this motion, self diffusion, is accessible to neutron studies. Actually, the most reliable "measurement" of it are not these neutron studies but computer studies [7,8] in which the average properties of a particle in the fluid are determined by computing the dynamical behavior of particles interacting by van der Waals forces. The quantity, $\omega \chi''(\omega)$, is exhibited for one value of the temperature and density in argon in Fig. 11. Also plotted [9] in Fig. 12 are the functions $\gamma'(\omega)$ and its Fourier transform $\bar{\gamma}'(t - t')$. Plotted for comparison on the same graphs are the Drude or Maxwell fit single collision time model, as obtained by Rice, [10] on the basis of rather more formidable arguments than Drude or Maxwell would ascribe to such an interpolation procedure.

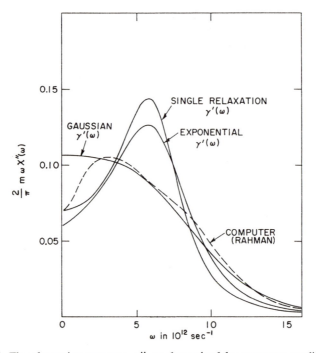

Fig. 11. The absorptive response, χ'', as determined by computer studies on liquid argon and various fits in terms of phenomenological laws described by simple functions $\gamma'(\omega)$.

A second physical example, involving only oscillators, is the most simplified version of the localised mode problem, a particle with a different mass but the same spring constant placed in a crystal which we idealize

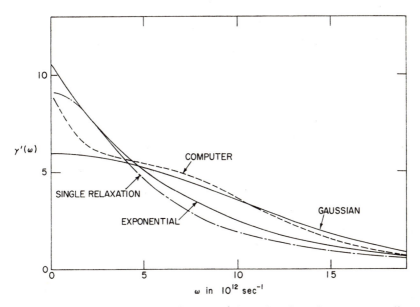

Fig. 12a. The actual phenomenological function, γ', determined from the computer studies and the fits.

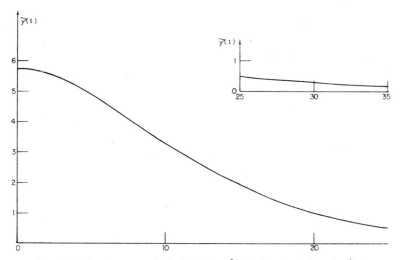

Fig. 12b. The phenomenological function $\tilde{\gamma}'(t)$ whose transform is $\gamma'(\omega)$.

by a linear chain.[11] The Hamiltonian for this system may be written as

$$H = \sum_{\alpha} \frac{p_\alpha^2}{2\bar{m}} + \frac{\bar{m}}{2} \sum \omega_{0\alpha\beta}^2 x_\alpha x_\beta - \frac{p_0^2}{2\bar{m}} \frac{\delta m}{m}; \qquad \delta m = m - \bar{m} \qquad (47)$$

where x_0 is the coordinate of the tagged particle whose mass is m. The equations of motion are

$$\bar{m}\ddot{x}_\alpha + \sum_\beta \bar{m}\omega_{0\alpha\beta}^2 x_\beta + \sum_\beta (\delta m)\,\delta_{\alpha 0}\delta_{\beta 0}\ddot{x}_\beta = 0. \tag{48}$$

By the same technique employed earlier we have

$$\sum_\beta (-\bar{m}z^2\delta_{\alpha\beta} + \bar{m}\omega_{0\alpha\beta}^2 - \delta m z^2\,\delta_{\alpha 0}\delta_{\beta 0})\,\chi_{\beta\gamma}(z) = \delta_{\alpha\gamma}. \tag{49}$$

This is a matrix equation for the matrix $\underline{\chi}$, whose inverse is

$$[\underline{\chi}^{-1}(z)]_{\alpha\beta} = -\bar{m}(z^2\delta_{\alpha\beta} - \omega_{0\alpha\beta}^2) - \delta m z^2\delta_{\alpha 0}\delta_{\beta 0}$$

$$= [\underline{\chi}^{-1}(z;\delta m = 0)]_{\alpha\beta} - \delta m z^2\delta_{\alpha 0}\delta_{\beta 0}. \tag{50}$$

Let us denote the matrix $\underline{\chi}(z;\delta m = 0)$ as $\underline{\chi}^0(z)$. Then we have

$$\chi_{\alpha\beta}^0(z) = \chi_{\alpha\beta}(z) - \delta m z^2 \underline{\chi}_{\alpha 0}^0(z)\,\chi_{0\beta}(z).$$

Since this equation implies that

$$\chi_{0\beta}(z) = \chi_{0\beta}^0(z) + \delta m z^2 \underline{\chi}_{00}^0(z)\,\chi_{0\beta}(z)$$

we have

$$\chi_{\alpha\beta}(z) = \chi_{\alpha\beta}^0(z) + \delta m z^2 \underline{\chi}_{\alpha 0}^0(z)\,[1 - \delta m z^2 \underline{\chi}_{00}^0(z)]^{-1}\,\chi_{0\beta}^0(z),$$

and in particular,

$$\chi_{00}(z) = \chi_{00}^0(z)\,[1 - \delta m z^2 \underline{\chi}_{00}^0(z)]^{-1}. \tag{51}$$

The quantity $\chi_{00}(z)$ is the correlation function $\chi(z)$ for the selected particle. We have for its correlation function

$$\chi^{-1}(z) = \chi^{0\,-1}(z) - \delta m z^2 \equiv -m[z^2 + iz\gamma(z)]. \tag{52}$$

Let us also introduce the symbol (in a notation which we will understand better a little later),

$$\bar{m}\chi_{vv}^0(z) \equiv \bar{m}z^2\chi^0(z) + 1;$$

$$\chi_{vv}^0(\omega) = \chi_{vv}^{0\,'}(\omega) + i\chi_{vv}^{0\,''}(\omega);$$

$$\chi_{vv}^{0\,''}(\omega) = \omega^2\chi^{0\,''}(\omega);\quad \chi_{vv}^{0\,'}(\omega) = P\int \frac{d\omega'}{\pi}\,\frac{\chi_{vv}^{0\,''}(\omega')}{\omega' - \omega}.$$

Then our equation for $\chi(z)$ reduces to

$$\delta m\chi(z) = \bar{m}\chi^0(z)\,[(\bar{m}/\delta m) + 1 - \bar{m}\chi_{vv}^0(z)]^{-1}. \tag{53}$$

In this equation all the dependence on m occurs through the explicit δm; the quantity $\bar{m}\chi_{vv}^0(z)$ is independent of m. The dependence on m can therefore be readily examined. A resonance in $\chi(z)$ will occur when the real part of the bracketed expression vanishes and the imaginary part is slowly varying over the width. The resonance will be infinitely sharp (a local mode) if the imaginary part vanishes where the real part does (which will be the case when m is sufficiently small). For larger m, but m which are still considerably smaller than $\bar{m}$ and again for $m \gg \bar{m}$, there will be a resonance, in the first case near the top of the band, in the second near the bottom. When $m \sim \bar{m}$ there is no resonance nor significant difference between χ'' and $\chi^{0''}$. In Fig. 13a the behavior of $\bar{m}\chi_{vv}^{0'}(\omega)$ and $\bar{m}\chi_{vv}^{0''}(\omega)$ is plotted together with

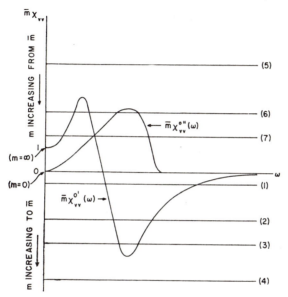

Fig. 13a. The spectrum of the perfect crystal. A plot of $\bar{m}\chi_{v}^{0'}(\omega)$ and $\bar{m}\chi_{vv}^{0''}(\omega)$. Also plotted are horizontal lines corresponding to different possible values for $[1 + \bar{m}/\delta m]$. Only with (1) does the intercept occur outside the continuum.

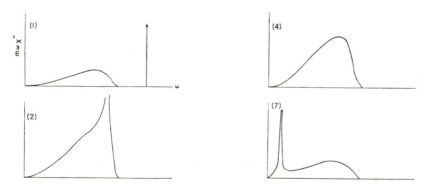

Fig. 13b. The spectrum of the particle of differing mass. In the situation, (1), a local mode occurs. When $m \cong \bar{m}$, (3) — (6) nothing dramatic happens. In other situations, 2) and 7), a resonance occurs. Qualitative graphs of $m\omega\chi''(\omega)$ corresponding to the various intercepts in Fig. 13a are shown.

$(\bar{m}/\delta m + 1)$ and in Fig. 13b the resultant $m\omega\chi''(\omega)$ is plotted for various values of m. The analytical expression these curves describe is

$$\chi''(\omega) = \frac{\bar{m}^2 \chi^{0''}(\omega)}{(\delta m)^2} \left\{ \left(1 + \frac{\bar{m}}{\delta m} - \bar{m}\chi_{vv}^{0'}(\omega)\right)^2 + \left(\bar{m}\chi_{vv}^{0''}(\omega)\right)^2 \right\}^{-1}. \quad (54)$$

B. Formal Development

Having illustrated various phenomena in term of this simple model, let us consider these arguments from a more general and rigorous point of view.[12] In this microscopic discussion, which we shall carry out quantum mechanically, the first and most important observation we wish to make concerns the rigorous identification of $\bar{\chi}''(t - t')$ with an equilibrium correlation function, a commutator (classically it would be a Poisson bracket). Having made this identification we may systematically examine its properties—notably symmetries, sum rules, and dispersion relations. We shall for example, see that the statement

$$\int \frac{d\omega}{\pi} \chi''(\omega)\,\omega = \frac{1}{m}$$

is just the statement

$$\left\langle \frac{i}{\hbar}\,[\dot{x}(t), x(t)] \right\rangle_{\text{eq.}} = \frac{i}{m\hbar}\left\langle [p(t), x(t)] \right\rangle_{\text{eq.}} = \frac{1}{m}.$$

From this microscopic viewpoint, the unproven statements we made earlier about function $\chi''(\omega)$ will emerge. In particular, we shall deduce the Nyquist theorem.

1. Time Dependent Perturbation Theory

We turn first to a general classical or quantum mechanical description of the effect of applying a weak external disturbance to a steady state. In mathematical terms, we suppose that prior to time t_0 the system is described by a density matrix, ϱ, which commutes with the time independent Hamiltonian H_0. Subsequent to t_0 an external disturbance is applied which couples to the observable properties, $A_j(\mathbf{r}t)$, of the system. We describe this disturbance by an additional term in the Hamiltonian

$$H_{\text{ext}}(t) = - \int d\mathbf{r} \sum_j A_j(\mathbf{r}t)\,a_j(\mathbf{r}t). \quad (1)$$

The functions $a_j(\mathbf{r}t)$ represent the generalized external forces. For example, the observables might include components of the magnetization, in which case the corresponding forces a_j would be the components of the external magnetic field. For our oscillator $A(\mathbf{r}t) \to x(t)$ and $a(\mathbf{r}t) \to F^{\text{ext}}(t)$. To calculate the expectation value at time t of the observable A_t we must calculate

$$Tr[\varrho A_i(\mathbf{r}t)] \equiv \langle A_i(\mathbf{r}t)\rangle_{\text{n.e.}} \tag{2}$$

where ϱ is the density matrix. Let us suppose for simplicity, that A does not depend on the time explicitly but only through the dynamical variables. We may either look upon (2) in the Heisenberg picture, in which the observable $A(\mathbf{r}t)$ changes in time because the dynamical variables on which it depends evolve and the density matrix is unaltered, or in the Schroedinger picture in which $A(\mathbf{r}t)$ is time independent but the dependence of the dynamical variables on t is accounted for by the time evolution of the density matrix. Independent of which picture we prefer we may write

$$\langle A_i(\mathbf{r}t)\rangle_{\text{n.e.}} = Tr[\varrho U^{-1}(tt_0)\, A_i(\mathbf{r}t_0)\, U(tt_0)] \tag{3}$$

where $U(tt_0)$ is the unitary operator which describes the way the system changes in time and satisfies the Schroedinger equation

$$i\hbar \frac{d}{dt} U(tt_0) = \big(H_0 + H_{\text{ext}}(t)\big)\, U(tt_0)$$

with the initial condition

$$U(t_0 t_0) = 1.$$

If we let

$$U(tt_0) = U_0(tt_0)\, U'(tt_0)$$

where U_0 satisfies $i\hbar \dfrac{d}{dt} U_0 = H_0 U_0$ we obtain

$$i\hbar \frac{d}{dt} U'(tt_0) = [U_0^{-1}(tt_0) H_{\text{ext}}(t) U_0(tt_0)]\, U'(tt_0)$$

$$i\hbar \frac{d}{dt} U'(tt_0) \equiv H^I_{\text{ext}}(t)\, U'(tt_0)$$

whose solution is

$$U'(tt_0) = 1 + \frac{1}{i\hbar}\int_{t_0}^{t} H^I_{\text{ext}}(t')\,dt' + \left(\frac{1}{i\hbar}\right)^2 \int_{t_0}^{t} H^I_{\text{ext}}(t') \int_{t_0}^{t} H^I_{\text{ext}}(t'')\eta(t'-t'') + \dots$$

$$\equiv T\left[\exp\left(\frac{1}{i\hbar}\int_{t_0}^{t} H^I_{\text{ext}}(t')\,dt'\right)\right]. \tag{4}$$

If we denote by $A^I(\mathbf{r}t)$, the Heisenberg operators for the Hamiltonian H_0, i.e.

$$A^I_i(\mathbf{r}t) = U_0^{-1}(tt_0)\, A_i(\mathbf{r}t_0)\, U_0(tt_0) \tag{5}$$

then we may write

$$Tr[\varrho U^{-1}A_i(\mathbf{r}t_0)\,U] = Tr[\varrho U'^{-1}A_i^I(\mathbf{r}t)\,U']$$

$$= Tr\left[\varrho\left\{A^I + \frac{i}{\hbar}\sum_j\int d\mathbf{r}'\int_{t_0}^t dt'\,[A_i^I(\mathbf{r}t),\,A_j^I(\mathbf{r}'t')]\,a_j(\mathbf{r}'t')\right\}\right] + \cdots$$

or, with the understanding that when there are no superscripts I, we are discussing the steady state density matrix commuting with H_0 and operators which evolve according to it,

$$\langle A_i(\mathbf{r}t)\rangle_{\text{n.e.}} \cong \langle A_i(\mathbf{r}t)\rangle + \frac{i}{\hbar}\sum_j\int d\mathbf{r}'\int_{t_0}^t dt'\,\langle[A_i(\mathbf{r}t),\,A_j(\mathbf{r}'t')]\rangle\,a_j(\mathbf{r}'t') + \cdots \tag{6}$$

We now define the absorptive response as the commutator

$$\tilde{\chi}_{ij}''(\mathbf{r}\mathbf{r}';t-t') \equiv \frac{1}{2\hbar}\langle[A_i(\mathbf{r}t),\,A_j(\mathbf{r}'t')]\rangle \tag{7}$$

$$= \int\frac{d\omega}{2\pi}\,e^{-i\omega(t-t')}\chi_{ij}''(\mathbf{r}\mathbf{r}';\omega). \tag{8}$$

In terms of χ'' and the step function $\eta(t-t')$ we may write

$$\delta\langle A_i(\mathbf{r}t)\rangle = \sum_j\int d\mathbf{r}'\int_{-\infty}^\infty 2dt'\,\tilde{\chi}_{ij}''(\mathbf{r}\mathbf{r}';t-t')\,a_j(\mathbf{r}'t')\,i\eta(t-t') \tag{9}$$

The corresponding classical expression is

$$\delta A_i(\mathbf{r}t) = -\sum_j\int_{t_0}^t \langle[A_i(\mathbf{r}t),\,A_j(\mathbf{r}'t')]_{\text{P.B.}}\rangle a_j(\mathbf{r}'t')$$

which leads us to define

$$\chi_{ij}''{}^{cl} = \frac{i}{2}\langle[A_i(\mathbf{r}t),\,A_j(\mathbf{r}'t')]_{\text{P.B.}}\rangle.$$

(Being more explicit requires unfortunately complicated notation, i.e.,

$$\chi_{ij}''{}^{cl} = \frac{i}{2}\left\langle\sum_\alpha\frac{\partial A_i(\mathbf{r}t)}{\partial\mathbf{r}_\alpha(t)}\frac{\partial A_j(\mathbf{r}'t')}{\partial\mathbf{p}_\alpha(t')} - \frac{\partial A_i(\mathbf{r}t)}{\partial\mathbf{p}_\alpha(t)}\frac{\partial A_j(\mathbf{r}'t')}{\partial\mathbf{r}_\alpha(t')}\right\rangle$$

with $\partial A_i(\mathbf{r};\mathbf{r}_\alpha(t),\,\mathbf{p}_\alpha(t))/\partial\mathbf{r}_\alpha(t) \equiv \partial A_i(\mathbf{r};e^{L_0(t-t')}\mathbf{r}_\alpha(t'),\,e^{L_0(t-t')}\mathbf{p}_\alpha(t'))/\partial\mathbf{r}_\alpha(t)$, the derivative at time t with respect to the dynamical variable into which the variable at the common time t' has evolved under the hamiltonian H_0. This value is classically determined by the Liouville operator L_0 associated with H_0 by the above expression.) To prove the classical version, we recall that if the system is described classically by a distribution function

$f = f(\mathbf{r}_\alpha(t), \mathbf{p}_\alpha(t))$, we may write to first order

$$f = f_0 + \sum_\alpha \left(\delta\mathbf{r}^\alpha(t) \frac{\partial}{\partial\mathbf{r}^\alpha(t)} + \delta\mathbf{p}^\alpha(t) \frac{\partial}{\partial\mathbf{p}^\alpha(t)} \right) f_0$$

$$\delta f = \sum_\alpha \left[\int_{t_0}^t dt'\, \delta v^\alpha(t') \frac{\partial}{\partial\mathbf{r}_0^\alpha(t)} + \int_{t_0}^t dt'\, F_{\text{ext}}^\alpha(t') \frac{\partial}{\partial\mathbf{p}_0^\alpha(t)} \right] f_0$$

$$= \int_{t_0}^t [f_0(t), H_{\text{ext}}(t')]_{\text{P.B.}}\, dt'.$$

This expression, like the one for the perturbed quantum mechanical density matrix, is only useful for calculating expectation values of operators for generic measurable quantities-functions of a few variables symmetrical in the many coordinates of the system and not distinguishing among them. For these quantities, we have

$$\delta\langle A_i(\mathbf{r}t)\rangle = \sum_j \int_{t_0}^t \langle [A_i(\mathbf{r}t), A_j(\mathbf{r}'t')]_{\text{P.B.}}\rangle\, a_j(\mathbf{r}'t')\, d\mathbf{r}'\, dt'$$

where as usual, the brackets indicate an ensemble average.

We now define

$$\tilde\chi_{ij}(\mathbf{r}\mathbf{r}'; t - t') = 2i\eta(t - t')\, \tilde\chi_{ij}''(\mathbf{r}\mathbf{r}'; t - t') \tag{10}$$

$$\delta\langle A_i(\mathbf{r}t)\rangle = \sum_j \int d\mathbf{r}' \int_{-\infty}^\infty dt'\, \tilde\chi_{ij}(\mathbf{r}\mathbf{r}'; t - t')\, a_j(\mathbf{r}'t'). \tag{9'}$$

The function $\tilde\chi_{ij}(\mathbf{r}\mathbf{r}'; t - t')$ is the Fourier transform of the complex response $\chi_{ij}(\mathbf{r}\mathbf{r}'; \omega)$. Moreover,

$$\chi_{ij}(\mathbf{r}\mathbf{r}'; \omega) = \chi_{ij}'(\mathbf{r}\mathbf{r}'; \omega) + \chi_{ij}''(\mathbf{r}\mathbf{r}'; \omega) \tag{11}$$

is the boundary value as z approaches ω on the real axis from above, of the analytic function of z

$$\chi_{ij}(\mathbf{r}\mathbf{r}'; z) = \int \frac{d\omega'}{\pi} \frac{\chi_{ij}''(\mathbf{r}\mathbf{r}'; \omega')}{\omega' - z}. \tag{12}$$

It follows immediately from these equations that χ' and χ'' satisfy Kramers-Kronig relations,

$$\chi_{ij}'(\mathbf{r}\mathbf{r}'; \omega) = P\int \frac{d\omega'}{\pi} \frac{\chi_{ij}''(\mathbf{r}\mathbf{r}'; \omega')}{\omega' - \omega}\,;\, \chi_{ij}''(\mathbf{r}\mathbf{r}'; \omega) = -P\int \frac{d\omega'}{\pi} \frac{\chi_{ij}'(\mathbf{r}\mathbf{r}'; \omega')}{\omega' - \omega}. \tag{13}$$

2. Symmetry Properties of the Response Function

(i) Since χ_{ij}'' is a commutator, it is antisymmetric under interchange of $\mathbf{r}$ with $\mathbf{r}'$, i with j, and t with t'. We therefore have

$$\tilde\chi_{ij}''(\mathbf{r}\mathbf{r}'; t - t') = -\tilde\chi_{ji}''(\mathbf{r}'\mathbf{r}; t' - t)$$

$$\chi_{ij}''(\mathbf{r}\mathbf{r}'; \omega) = -\chi_{ji}''(\mathbf{r}'\mathbf{r}; -\omega). \tag{14}$$

(ii) The fact that $\tilde{\chi}''_{ii}$ is the commutator of Hermitian operators leads to the identity

$$[\tilde{\chi}''_{ij}(\mathbf{rr}'; t - t')]^* = -\tilde{\chi}''_{ij}(\mathbf{rr}'; t - t'), \quad \text{i.e., } \tilde{\chi}''_{ij} \text{ is imaginary,}$$

$$[\chi''_{ij}(\mathbf{rr}'; \omega)]^* = -\chi''_{ij}(\mathbf{rr}'; -\omega). \tag{15}$$

(iii) The effect of (i) and (ii), i.e., transposition and complex conjugation, implies that $\tilde{\chi}''$ is hermitian, i.e.

$$[\tilde{\chi}''_{ij}(\mathbf{rr}'; t - t')]^* = \tilde{\chi}''_{ji}(\mathbf{r'r}; t' - t)$$

$$[\chi''_{ij}(\mathbf{rr}'; \omega)]^* = \chi''_{ji}(\mathbf{r'r}; \omega). \tag{16}$$

The part of $\chi''_{ij}(\mathbf{rr};' \omega)$ which is symmetric under interchange of i with j and $\mathbf{r}$ with $\mathbf{r}'$ is both real and odd in ω while the antisymmetric part is imaginary and even in ω. These statements imply in particular that if $\chi''_{ii}(\mathbf{rr}'; \omega)$ is spatially invariant $(\chi''_{ii} = \chi''_{ii}(|\mathbf{r} - \mathbf{r}'|; \omega))$ it is real and odd in the frequency.

(iv) In general $\chi''_{ij}(\mathbf{rr}'; \omega)$ need not be real. More particularly its reality properties are connected with time reversal. The effect of the antiunitary operation, θ, on states of the system is to transform scalar products according to $\langle \theta\alpha | \theta\beta \rangle = \langle \beta | \alpha \rangle$. Corresponding to this transformation on the states one has the similarity transformation on the operators $A_i \to A'_i = \theta A_i \theta^{-1}$ and the composition law $A_i A_j \to (A_i A_j)' = A'_i A'_j$. Moreover for observables $A_i(\mathbf{r}t)$, (which are described by Hermitian operators) the effect of applying the time reversal operator is to give another hermitian operator which usually will have a definite signature ε_i

$$A'_i(\mathbf{r}t) = \theta A_i(\mathbf{r}t) \theta^{-1} = \varepsilon_i A_i(\mathbf{r} - t)$$

(e.g., $\varepsilon = +1$ for position, electric fields; $\varepsilon = -1$ for velocities and magnetic fields). We will then have

$$(\theta[A_i(\mathbf{r}t), A_j(\mathbf{r't'})] \theta^{-1})^\dagger = [\theta A_j(\mathbf{r't'}) \theta^{-1}, \theta A_i(\mathbf{r}t) \theta^{-1}]$$

$$= -\varepsilon_i \varepsilon_j [A_i(\mathbf{r} - t), A_j(\mathbf{r'} - t')].$$

Consequently, whenever the Hamiltonian and the ensemble of states are invariant under time reversal, since $\langle \alpha | B | \alpha \rangle = \langle \theta\alpha | (\theta B \theta^{-1})^\dagger | \theta\alpha \rangle$,

$$\tilde{\chi}''_{ij}(\mathbf{rr}'; t - t') = -\varepsilon_i \varepsilon_j \tilde{\chi}''_{ij}(\mathbf{rr}'; t' - t) = \varepsilon_i \varepsilon_j \tilde{\chi}''_{ji}(\mathbf{r'r}; t - t'),$$

$$\chi''_{ij}(\mathbf{rr}'; \omega) = -\varepsilon_j \varepsilon_i \chi''_{ij}(\mathbf{rr}'; -\omega) = \varepsilon_i \varepsilon_j \chi''_{ji}(\mathbf{r'r}; \omega). \tag{17}$$

This means that if A_i and A_j have the same signature under time reversal $\chi''_{ii}(\mathbf{rr}'; \omega)$ is odd in ω, real, and symmetric under interchange of i with j and $\mathbf{r}$ with $\mathbf{r}'$. If they have opposite signature, $\chi''_{ij}(\mathbf{rr}'; \omega)$ is even, imaginary and antisymmetric.

If the hamiltonian and ensemble involve a magnetic field or some other property which changes sign under time reversal, then the more complicated relation

$$\chi_{ij}''(\mathbf{rr}';\omega;\mathbf{B}) = \varepsilon_i \varepsilon_j \chi_{ji}''(\mathbf{r'r};\omega;-\mathbf{B})$$
$$= -\varepsilon_i \varepsilon_j \chi_{ij}''(\mathbf{rr}';-\omega;-\mathbf{B}) \qquad (18)$$

is obtained because the density matrix of time reversed states is different. As a result, for two operators with the same signature under time reversal there will be an additional part of $\chi_{ij}''(\mathbf{rr}';\omega)$ which is odd in the field, $\mathbf{B}$, even in ω, imaginary, and antisymmetric in i, $\mathbf{r}$ and j, $\mathbf{r}'$.

The symmetry properties of $\chi_{ij}'(\mathbf{rr}';\omega, \mathbf{B})$ are determined from the relation

$$\chi_{ij}'(\mathbf{rr}';\omega;\mathbf{B}) = P \int \frac{d\omega'}{\pi} \frac{\chi_{ij}''(\mathbf{rr}';\omega';\mathbf{B})(\omega'+\omega)}{\omega'^2 - \omega^2} \qquad (19)$$

which means that they are identical apart from the interchange of evenness and oddness in ω. (For example, in our illustration, a Lorentz Force would give rise to a term $\varepsilon_{ijk} i\omega B_k$ in χ_{ij}'.)

3. Identification of χ'' with Dissipation

We may identify the work done with the explicit rate of change of the hamiltonian and consequently associate the matrix $\chi_{ij}''(\mathbf{rr}';\omega)$ with dissipation. In this case, there is not necessarily any reality property for the off diagonal elements, or signature for the individual components. We have, however, the statement that the rate of change of energy or the rate at which work is done on the system is given by the explicit rate of change of the expectation value of the hamiltonian

$$-\frac{dW}{dt} = \sum_i \int \langle A_i(\mathbf{r}t)\rangle_{\text{n.e.}} \dot{a}_i(\mathbf{r}t)\, d\mathbf{r}$$

$$= \sum_i \int \langle A_i(\mathbf{r}t)\rangle_{\text{eq.}} \dot{a}_i(\mathbf{r}t)\, d\mathbf{r} + \sum_{ij} \int d\mathbf{r} \int d\mathbf{r}' \int_{-\infty}^{\infty} dt'\, \dot{a}_i(\mathbf{r}t)\, \tilde{\chi}_{ij} \,(\mathbf{rr}';t-t')a_j(\mathbf{r}'t')$$

$$+ \text{ (terms of order } a^3). \qquad (20)$$

Thus the mean rate of change of energy in a monochromatic external field

$$a_i(\mathbf{r}t) = \text{Re } a_i(\mathbf{r}) e^{-i\omega t} = \tfrac{1}{2}[a_i(\mathbf{r}) e^{-i\omega t} + a_i^*(\mathbf{r}) e^{i\omega t}]$$

is given by

$$-\frac{\overline{dW}}{dt} = \sum_{ij} \left\{ \frac{i\omega}{4} \int d\mathbf{r} \int d\mathbf{r}' a_i^*(\mathbf{r})\, \chi_{ij}(\mathbf{rr}';\omega)\, a_j(\mathbf{r}') \right.$$

$$\left. - \frac{i\omega}{4} \int d\mathbf{r} \int d\mathbf{r}' \, a_i(\mathbf{r})\, \chi_{ij}\,(\mathbf{rr}';-\omega)a_j^*(\mathbf{r}') \right\} \qquad (21)$$

6

which, in view of the symmetries $\chi_{ij}''(\mathbf{rr}';\omega) = -\chi_{ji}''(\mathbf{r'r};-\omega)$ and $\chi_{ij}'(\mathbf{rr}';\omega) = \chi_{ji}'(\mathbf{r'r};-\omega)$, becomes

$$\frac{\overline{dW}}{dt} = \frac{1}{2}\sum_{ij}\omega\int d\mathbf{r}\int d\mathbf{r}'a_i^*(\mathbf{r})\,\chi_{ij}''(\mathbf{rr}';\omega)\,a_j(\mathbf{r}'). \tag{22}$$

For plane waves, $a_i(\mathbf{r}) = a_i e^{i\mathbf{k}\cdot\mathbf{r}}$, and translationally invariant systems, we have per unit volume and time

$$\frac{1}{V}\frac{\overline{dW}}{dt} = \frac{1}{2}\sum_{ij}a_i^*\chi_{ij}''(\mathbf{k}\omega)\,\omega a_j$$

with

$$\chi_{ij}''(\mathbf{rr}';\omega) = \int\frac{d\mathbf{k}}{(2\pi)^3}\,e^{i\mathbf{k}\cdot(\mathbf{r}-\mathbf{r}')}\chi_{ij}''(\mathbf{k}\omega). \tag{23}$$

For an arbitrary disturbance which vanishes initially and finally the total work done is given by

$$W = \sum_{ij}\int d\mathbf{r}\int dt\int d\mathbf{r}'\int dt'a_i(\mathbf{r}t)\frac{\partial}{\partial t}\tilde{\chi}_{ij}''(\mathbf{rr}';t-t')\,a_j(\mathbf{r}'t') \tag{24}$$

in view of the antisymmetry of $\partial\tilde{\chi}_{ij}'(\mathbf{rr}';t-t')/\partial t$.

The expressions for a monochromatic field may be cast in an alternative form which is perhaps more familiar by noting that they are equivalent to the golden rule. Specifically, by introducing an intermediate set of states labelled by E' and other quantum numbers ξ', we obtain

$$\frac{\overline{dW}}{dt} = \sum_{ij}\int d\mathbf{r}\int d\mathbf{r}'a_i^*(\mathbf{r})\frac{\omega}{2}\chi_{ij}''(\mathbf{rr}';\omega)\,a_j(\mathbf{r}')$$

$$= \int\frac{d(t-t')}{\hbar}e^{i\omega(t-t')}\omega\sum_{\substack{\xi'E'\\\xi E}}w_{\xi E}\left\{\langle E\xi|\frac{1}{2}\sum_i\int d\mathbf{r}a_i^*(\mathbf{r})\,A_i(\mathbf{r})\,|E'\xi'\rangle\,e^{\frac{i}{\hbar}(E-E')(t-t')}\right.$$

$$\times\,\langle E'\xi'|\frac{1}{2}\sum_j\int d\mathbf{r}'a_j(\mathbf{r}')\,A_j(\mathbf{r}')\,|E\xi\rangle - \langle E\xi|\frac{1}{2}\sum_j\int d\mathbf{r}'a_j(\mathbf{r}')\,A_j(\mathbf{r}')\,|E'\xi'\rangle$$

$$\times\,e^{-\frac{i}{\hbar}(E-E')(t-t')}\left.\langle E'\xi'|\frac{1}{2}\sum_i\int d\mathbf{r}a_i^*(\mathbf{r})\,A_i(\mathbf{r})\,|E\xi\rangle\right\}$$

$$= \sum_{\xi E}w_{\xi E}\hbar\omega[P_{\xi E}(\hbar\omega) - P_{\xi E}(-\hbar\omega)] \tag{25}$$

where $w_{\xi E}$ is the weight of the state ξE in the ensemble, ϱ_{eq}.

$$P_{\xi E}(\hbar\omega) = \frac{2\pi}{\hbar}\sum_{\xi'}\left|\langle E\xi|\frac{1}{2}\sum_i\int d\mathbf{r}a_i^*(\mathbf{r})\,A_i(\mathbf{r})\,|E+\hbar\omega\xi'\rangle\right|^2$$

is the probability for absorption of energy $\hbar\omega$ due to the presence of the interaction hamiltonian, and $P_{\xi E}(-\hbar\omega)$ the corresponding induced emission probability. One might prefer to read this discussion in reverse considering the result more familiar than the starting point.

Unless the system is an amplifier and certainly for a system in thermal equilibrium, we must have dissipation at each frequency,

$$\omega \chi_{ii}''(\mathbf{rr}; \omega) \geqslant 0.$$

From the Kramers-Kronig relation it then follows that the matrix χ' is positive definite for low frequencies and negative definite at high frequencies. This corresponds to the statement for a forced oscillator that at low frequencies its displacement will be in the direction of the force, and therefore non-dissipative, and at the high frequencies its response will be 180° out of phase and also non-dissipative. Dissipation, not amplification, occurs for intermediate frequencies, and is peaked at the oscillator frequency, where resonance occurs.

4. Sum Rules or Moment Expansions

The short time or high frequency behavior of the correlation function may be characterized in terms of the time derivatives or frequency moments of the correlation function, that is, in terms of the quantities

$$\frac{1}{\hbar} \left\langle \left[\left(i\frac{d}{dt} \right)^n A_i(\mathbf{r}t), A_j(\mathbf{r}'t') \right] \right\rangle = \int \frac{d\omega}{\pi} \omega^n \chi_{ij}''(\mathbf{rr}'; \omega)$$

or

$$\frac{1}{\hbar} \left\langle \left[\left[\left[A_i(\mathbf{r}t), \frac{H}{\hbar} \right], \frac{H}{\hbar} \right] .. \right], A_j(\mathbf{r}'t') \right\rangle = \int \frac{d\omega}{\pi} \omega^n \chi_{ij}''(\mathbf{rr}'; \omega). \qquad (28)$$

These expressions are known as moment sum rules, and the left hand side, a multiple commutator with the hamiltonian, may in some instances be exactly evaluated. The resulting expansions for the correlation function, however,

$$\chi_{ij}(\mathbf{rr}'; z) = \int \frac{d\omega}{\pi} \frac{\chi_{ij}''(\mathbf{rr}'; \omega)}{\omega - z} = -\sum_{p=1}^{\infty} \frac{\langle \omega_{ij}^p(\mathbf{rr}') \rangle}{z^p} \chi_{ij}(\mathbf{rr}'; 0)$$

with

$$\langle \omega_{ij}^p(\mathbf{rr}') \rangle \chi_{ij}(\mathbf{rr}'; 0) \equiv \int \frac{d\omega}{\pi} \left[\frac{\chi_{ij}''(\mathbf{rr}'; \omega)}{\omega} \right] \omega^p \qquad (29)$$

are only asymptotic. They hold rigorously only for frequencies higher than any characteristic frequency ω of the system, and as they predict that $\chi_{ij}'(\mathbf{rr}'; \omega)$ is negative, they are sensible asymptotically only when ω is large compared to all important frequency contributions to χ''.

The simplest illustration of these sum rules is the one we cited for the oscillator

$$\frac{i}{\hbar} \langle [\dot{x}(t), x(t)] \rangle = \int \frac{d\omega}{\pi} \omega \chi_{xx}''(\omega) \qquad (30)$$

6*

which, for velocity independent forces, gives

$$\int \frac{d\omega}{\pi} \omega \chi_{xx}''(\omega) = \frac{1}{m} \left\langle \left[\frac{i}{\hbar} p(t), x(t) \right] \right\rangle = \frac{1}{m} \equiv \langle \omega_{xx}^2 \rangle \chi_{xx}(0). \quad (31)$$

5. Fluctuation Dissipation Theorems

For most situations the stationary ensemble which characterizes the system is canonical, or in other words

$$w_{\xi E} = e^{-\beta E} / \sum_{\xi E} e^{-\beta E}; \quad \varrho = e^{-\beta H} [Tr e^{-\beta H}]^{-1} \quad (32)$$

where $\beta = (kT)^{-1}$. The time translation property of the weighting factor for such a canonical ensemble and the cyclical property of the trace imply the identities

$$Tr[e^{-\beta H} A_i(\mathbf{r}t) A_j(\mathbf{r}'t')] = Tr[A_i(\mathbf{r}t + i\beta\hbar) e^{-\beta H} A_j(\mathbf{r}'t')]$$

$$= Tr[e^{-\beta H} A_j(\mathbf{r}'t') A_i(\mathbf{r}t + i\beta\hbar)]. \quad (33)$$

Moreover, $Tr[\exp(-\beta H) A(\mathbf{r}t)]$ is independent of time. Consequently provided the time Fourier transform

$$\langle (A_i(\mathbf{r}t) - \langle A_i(\mathbf{r}t) \rangle) (A_j(\mathbf{r}'t') - \langle A_j(\mathbf{r}'t') \rangle) \rangle$$

$$\equiv \tilde{S}_{ij}(\mathbf{r}\mathbf{r}'; t - t') = \int \frac{d\omega}{2\pi} S_{ij}(\mathbf{r}\mathbf{r}'; \omega) e^{-i\omega(t-t')} \quad (34)$$

exists (and it will in a sufficiently specified ensemble), it satisfies

$$S_{ij}(\mathbf{r}\mathbf{r}'; \omega) = S_{ji}(\mathbf{r}'\mathbf{r}; -\omega) e^{\beta\omega\hbar}$$

and therefore

$$\chi_{ij}''(\mathbf{r}\mathbf{r}'; \omega) = \frac{1}{2\hbar} (1 - e^{-\beta\omega\hbar}) S_{ij}(\mathbf{r}\mathbf{r}'; \omega)$$

$$= \frac{1}{2\hbar} (e^{\beta\omega\hbar} - 1) S_{ji}(\mathbf{r}'\mathbf{r}; -\omega). \quad (35)$$

Likewise the transform of the symmetrized product

$$\tfrac{1}{2} \langle \{[A_i(\mathbf{r}t) - \langle A_i(\mathbf{r}t) \rangle], [A_j(\mathbf{r}'t') - \langle A_j(\mathbf{r}'t') \rangle]\} \rangle$$

$$\equiv \tilde{\varphi}_{ij}(\mathbf{r}\mathbf{r}'; t - t') = \int \frac{d\omega}{2\pi} \varphi_{ij}(\mathbf{r}\mathbf{r}'; \omega) e^{-i\omega(t-t')} \quad (36)$$

satisfies the identity

$$\varphi_{ij}(\mathbf{r}\mathbf{r}'; \omega) = \tfrac{1}{2}(1 + e^{-\beta\omega\hbar}) S_{ij}(\mathbf{r}\mathbf{r}'; \omega) \quad (37)$$

and the fluctuation dissipation theorem in the form

$$\frac{1}{2}\,\varphi_{ij}(\mathbf{rr}';\omega) = \hbar\omega\left[\frac{1}{2} + \frac{1}{e^{\beta\omega\hbar} - 1}\right]\frac{\chi_{ij}''(\mathbf{rr}';\omega)}{\omega}$$

$$= \frac{\hbar}{2}\coth\frac{\beta\omega\hbar}{2}\,\chi_{ij}''(\mathbf{rr}';\omega). \tag{38}$$

It is the failure of the fluctuation dissipation theorem in a canonical or grand canonical ensemble, when that ensemble is insufficient because there are order parameters, which provides the weak link in the apparent proof we shall give that fluid hydrodynamic equations are always correct. Its validity in more restrictive ensembles with specified order parameters leads to the appropriate hydrodynamic equations for these systems.

Classically we may arrive at an analogous result by partially integrating with respect to $\mathbf{p}_\alpha$ and $\mathbf{r}_\alpha$

$$-\sum_\alpha \int \pi d\mathbf{p}_\alpha^0(t')\,d\mathbf{r}_\alpha^0(t')\left[\frac{\partial}{\partial \mathbf{r}_\alpha^0(t)}\,A_i(\mathbf{rt})\frac{\partial}{\partial \mathbf{p}_\alpha^0(t')}\,A_j(\mathbf{r}'t')\right.$$

$$\left. - \frac{\partial}{\partial \mathbf{p}_\alpha^0(t)}\,A_i(\mathbf{rt})\frac{\partial}{\partial \mathbf{r}_\alpha^0(t')}\,A_j(\mathbf{r}'t')\right]e^{-\beta H_0(\mathbf{r}_\alpha^0\mathbf{p}_\alpha^0)}.$$

We thereby obtain

$$2i\tilde{\chi}_{ij}''^{cl}(\mathbf{rr}';t - t') = -\beta\frac{\partial}{\partial t}\tilde{S}_{ij}^{cl}(\mathbf{rr}';t - t') = -\beta\frac{\partial}{\partial t}\tilde{\varphi}_{ij}^{cl}(\mathbf{rr}';t - t')$$

$$\chi_{ij}''^{cl}(\mathbf{rr}';\omega) = \frac{\beta\omega}{2}\,S_{ij}^{cl}(\mathbf{rr}';\omega) = \frac{\beta\omega}{2}\,\varphi_{ij}^{cl}(\mathbf{rr}';\omega)$$

in accordance with the classical limit of the quantum mechanical result.

6. Dispersion Relation for Response Function in Time Reversal Invariant Systems

In our previous discussion we used a dispersion relation[13] which replaced the unknown $\chi(z)$ by the unknown $\gamma(z)$. To prove that this is possible it is sufficient to note that for complex z, $\chi(z) \neq 0$ (because $\mathrm{Im}\, z\, \chi(z) \neq 0$ whenever $\omega\chi''(\omega) > 0$). It then follows that $\chi^{-1}(z)$ is analytic, and consequently that the remainder we obtain when we subtract its leading term for large z

$$\chi^{-1}(z) + [z^2/\langle\omega^2\rangle]\,\chi^{-1}(0)$$

is analytic and approaches a constant at infinity. We may therefore write the difference either as

$$[-iz\,\hat{\Gamma}(z)/\langle\omega^2\rangle]\,\chi^{-1}(0) \quad \text{or} \quad [(-iz\,\hat{\gamma}(z) + \omega_0^2)/\langle\omega^2\rangle]\,\chi^{-1}(0) \tag{39}$$

where $\hat{\Gamma}$ and $\hat{\gamma}$ are analytic except on the real axis and approach zero as $z \to \infty$. In the latter expression ω_0^2 may be chosen so that $z\hat{\gamma}(z)$ approaches zero as $z \to 0$.

For our further work, it will be useful to prove another dispersion relation[14] which is related to the one above. This second dispersion relation depends on the fact that $\chi(z) \neq \chi(0)$ when $\mathrm{Im}\, z \neq 0$; it is proven by showing that $\mathrm{Im}\,([\chi(z) - \chi(0)]/z) \neq 0$. We may therefore conclude that $[-1 + \chi(0)/\chi(z)]^{-1}$ is analytic and vanishes at infinity. If we call it $i\Gamma(z)/z$, we may write

$$-1 + \frac{\chi(0)}{\chi(z)} = \frac{z}{i\Gamma(z)}$$

or

$$-\frac{\chi(z)}{\chi(0)} + 1 = \frac{z}{z + i\Gamma(z)} = \left[1 + \int \frac{d\omega}{\pi} \frac{\Gamma'(\omega)}{\omega^2 - z^2}\right]^{-1} \tag{40}$$

with

$$\Gamma(z) = \int \frac{d\omega}{\pi i} \frac{\Gamma'(\omega)}{\omega - z} = \int \frac{d\omega}{\pi i} \frac{\Gamma'(\omega)\, z}{\omega^2 - z^2}. \tag{41}$$

As z approaches the real axis, $\Gamma(z) \to \Gamma'(\omega) + i\Gamma''(\omega)$ with

$$\Gamma''(\omega) = -P \int \frac{d\omega'}{\pi} \frac{\Gamma'(\omega')\,\omega}{\omega'^2 - \omega^2}. \tag{42}$$

We may rewrite these relations in the form

$$-\chi(z) + \chi(0) = \frac{z^2 \chi(0)}{z^2 + zi\Gamma(z)} \tag{43}$$

$$\chi''(\omega) = \frac{\omega^2 \chi(0)\, \Gamma'(\omega)}{\left(\omega^2 - \omega\Gamma''(\omega)\right)^2 + \left(\omega\Gamma'(\omega)\right)^2}. \tag{44}$$

Let us again separate out any term in $-iz\Gamma(z)$ which behaves as a constant, ω_0^2, near $z = 0$, or equivalently, any term in $\Gamma'(\omega)$ which behaves as $\pi\omega_0^2\,\delta(\omega)$ writing

$$-iz\Gamma(z) = \omega_0^2 - iz\gamma(z)$$

$$\Gamma'(\omega) = \pi\omega_0^2\,\delta(\omega) + \gamma'(\omega). \tag{45}$$

Then we have (if there is such a contribution)

$$-\chi(z) + \chi(0) = \frac{z^2 \chi(0)}{z^2 - \omega_0^2 + iz\gamma(z)} \tag{46}$$

$$\frac{\chi''(\omega)}{\omega} = \frac{\omega^3 \chi(0)\, \gamma'(\omega)}{\left(\omega^2 - \omega_0^2 - \omega\gamma''(\omega)\right)^2 + \left(\omega\gamma'(\omega)\right)^2}. \tag{47}$$

We see, in other words, that a δ-function contribution of $\gamma'(\omega)$ leads to oscillatory behavior at low frequencies and its absence ($\omega_0 = 0$) to relaxation at low frequencies or long times.

Now suppose the correlation function in this section is the velocity auto-correlation function, $\chi_{vv}(z)$ for the system discussed in Section A. We then have the identities

$$m\chi_{vv}(0) = 1; \chi_{vv}''(\omega) = \omega^2\chi_{xx}''(\omega) \tag{48}$$

and also

$$m\chi_{vv}(z) = 1 + mz^2\chi_{xx}(z); \tag{49}$$

where χ_{xx} is the displacement auto-correlation function discussed in that section. Then the two equations derived above are equivalent to ($\gamma(z) = \hat{\gamma}(z)$) and

$$\chi_{xx}(z) = \frac{1}{-m[z^2 - \omega_0^2 + iz\gamma(z)]} \tag{50}$$

and

$$\omega\chi_{xx}''(\omega) = \frac{\omega^2\gamma'(\omega)}{m[(\omega^2 - \omega_0^2 - \omega\gamma''(\omega))^2 + (\omega\gamma'(\omega))^2]}. \tag{51}$$

In other words, the form we introduced in our preliminary discussion (eq. A41) is the most general expression for the displacement correlation function for a quantum mechanical particle, in a medium which is time reversal invariant.

Next let us recall that our expression of the sum rule implies that as $z \to \infty$

$$\frac{i}{\hbar}\langle[\ddot{x}, \dot{x}]\rangle = \int\frac{d\omega}{\pi}\omega^3\chi_{xx}''(\omega) = \int\frac{d\omega}{\pi}\omega\chi_{vv}''(\omega) \equiv \frac{\langle\omega_{vv}^2\rangle}{m} \equiv \langle\omega_{xx}^4\rangle\chi_{xx}(0)$$

$$= \frac{1}{m}\int\frac{d\omega}{\pi}\Gamma'(\omega) = \frac{\omega_0^2}{m} + \frac{1}{m}\int\frac{d\omega}{\pi}\gamma'(\omega) \equiv \frac{\omega_\infty^2}{m} \tag{52}$$

so that

$$\int\frac{d\omega}{\pi}\gamma'(\omega) = \omega_\infty^2 - \omega_0^2.$$

Our adhoc phenomenological description, then, was the simplest description consistent with these properties and amounted to taking ω_0 to be the oscillator spring constant and

$$\gamma'(\omega) = (\omega_\infty^2 - \omega_0^2)\frac{\tau}{1 + (\omega\tau)^2}. \tag{54}$$

As the subsequent discussion showed there was no particularly significant reason for choosing this one parameter fit to $\gamma'(\omega)$, and indeed, the true $\gamma'(\omega)$ must decay more rapidly, satisfying, for example,

$$\int\frac{d\omega}{\pi}\omega^5\chi_{xx}''(\omega) = \frac{1}{m}\left\{\int\frac{d\omega}{\pi}\Gamma'(\omega)\omega^2 + \left(\int\frac{d\omega}{\pi}\Gamma'(\omega)\right)^2\right\} \equiv \frac{\langle\omega_{vv}^4\rangle}{m}$$

$$\equiv \langle\omega_{xx}^6\rangle\chi_{xx}(0). \tag{55}$$

Indeed Rahman and Nijboer[15] have computed not only $\langle \omega_{vv}^2 \rangle$ but $\langle \omega_{vv}^4 \rangle$. These two expressions serve to determine the two parameters $\gamma'(0)$ and τ of a fit in which one takes $\gamma'(\omega) = \gamma_E'(0) \exp(-2|\omega| \tau_E/\pi)$, or $\gamma'(\omega) = \gamma_G'(0) \exp(-\omega^2 \tau_G^2/\pi)$. We have

$$\gamma_E'(0) = \langle \omega_{vv}^2 \rangle \tau_E \quad \text{and} \quad \tau_E^2 = \frac{\pi^2}{2} \frac{\langle \omega_{vv}^2 \rangle}{\langle \omega_{vv}^4 \rangle - \langle \omega_{vv}^2 \rangle^2}$$

or

$$\gamma_G'(0) = \langle \omega_{vv}^2 \rangle \tau_G \quad \text{and} \quad \tau_G^2 = \frac{\pi}{2} \frac{\langle \omega_{vv}^2 \rangle}{\langle \omega_{vv}^4 \rangle - \langle \omega_{vv}^2 \rangle^2} \tag{56}$$

The resultant curve is compared with the "experimental" expression in Fig. 12. For the exponential form the computed value of the diffusion constant, with no adjustable parameter, $D \equiv [\beta m \gamma_E'(0)]^{-1}$, is about 15 percent too low. (Note that this is *not* the Kirkwood prescription[16] for the diffusion constant. His prescription does not involve the fourth moment but the identification

$$\gamma_K'(0) = \langle \omega_{vv}^2 \rangle^{\frac{1}{2}}.$$

The ratio of γ's in the Drude and Kirkwood prescriptions is therefore given by

$$\gamma_E'/\gamma_K' = \sqrt{\pi} \, \gamma_G'/\gamma_K' = \frac{\pi}{\sqrt{2'}} \frac{\langle \omega_{vv}^2 \rangle}{(\langle \omega_{vv}^4 \rangle - \langle \omega_{vv}^2 \rangle^2)^{\frac{1}{2}}} \, ;$$

in charged systems the Drude expression for the conductivity reduces to $\omega_p^2 \tau$ whereas the Kirkwood expression would be ω_p, with ω_p the plasma frequency.)

C. Hydrodynamic Examples

1. Spin Magnetization

In our discussion thus far we have not put any restrictions on the observables A_i. Indeed, in our oscillator example, we have taken A_i to be the coordinate or velocity of a single particle. For certain variables, however, we can make some further statements. These are the variables which are associated with the densities of conserved quantities. Because the integrals of these densities are time independent, the Fourier transforms in time of their spatial integrals (i.e. the $\mathbf{k} = 0$ part of their spatial Fourier transforms) are proportional to $\delta(\omega)$. Correspondingly, when these conserved quantities vary slowly in space, they will be almost time independent. Their behavior under these slowly varying conditions is governed by the laws of hydrodynamics. For small deviations, linearized hydrodynamics is sufficient. But their behavior is also described by the correlation functions. For this reason[17] we may deduce the proper equations of linearized hydrodynamics, and the associated transport coefficients, if we can calculate (theoretically) or measure (experimentally) the long wavelength low frequency correlation functions. (See Fig. 14).

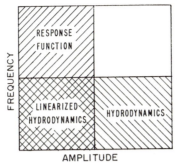

Fig. 14. Logical diagram indicating connection between response function theory which is valid for small amplitude disturbances, and hydrodynamic theory which is valid at long wavelengths and low frequencies. The intersection, linearized hydrodynamics, can be described in either fashion.

In order to fix our attention on a specific example, let us consider briefly the case in which we have a magnetization which can be modified by an external field. Indeed, suppose that the system is one, like He³ in which the magnetization is not an important part of the interparticle force and in which there are no spin orbit forces so that the magnetization satisfies a conservation law

$$\frac{\partial M(\mathbf{r}t)}{\partial t} + \nabla \cdot \mathbf{J}^M(\mathbf{r}t) = 0 \tag{1}$$

with $M(\mathbf{r}t) = \mu[n_+(\mathbf{r}t) - n_-(\mathbf{r}t)]$ and $\mathbf{J}^M(\mathbf{r}t) = \mu[\mathbf{j}_+(\mathbf{r}t) - \mathbf{j}_-(\mathbf{r}t)]$. In this example μ is the magnetic moment, $n_\pm$ the density of particles with spin in two opposite directions and $\mathbf{j}_\pm$ the corresponding particle currents. Let us calculate at time $t = 0$ and subsequently the effect of applying an external disturbance, described by

$$H = -\int d\mathbf{r} M(\mathbf{r}t)\, B^{\text{ext}}(\mathbf{r})\, e^{\varepsilon t} \eta(-t). \tag{2}$$

First, we note that if $B^{\text{ext}}(\mathbf{r}) = B_0 e^{i\mathbf{k}\cdot\mathbf{r}}$ we have

$$\delta\langle M(\mathbf{r}0)\rangle = \int_{-\infty}^{0} dt'\, \tilde{\chi}_{MM}(k;-t')\, e^{\varepsilon t'}\, B^{\text{ext}}(\mathbf{r}) \tag{3}$$

$$= \int_{-\infty}^{0} dt' \int_{-\infty}^{\infty} \frac{d\omega}{\pi}\, i\chi''_{MM}(k\omega)\, e^{i\omega t'} e^{\varepsilon t'}\, B^{\text{ext}}(\mathbf{r}) = \int_{-\infty}^{\infty} \frac{d\omega}{\pi}\, \frac{\chi''_{MM}(k\omega)}{\omega - i\varepsilon}\, B^{\text{ext}}(\mathbf{r})$$

$$= \chi_{MM}(k\, i\varepsilon)\, B^{\text{ext}}(\mathbf{r}). \tag{4}$$

Next we note that

$$\delta\langle M(\mathbf{r}t)\rangle = \int_{-\infty}^{0} dt' \int_{-\infty}^{\infty} \frac{d\omega}{\pi}\, i\chi''_{MM}(k\omega)\, e^{-i\omega(t-t')} e^{\varepsilon t'} B^{\text{ext}}(\mathbf{r}) \tag{5}$$

and that

$$M(\mathbf{r}z) \equiv \int_{0}^{\infty} dt\; e^{izt}\, \delta\langle M(\mathbf{r}t)\rangle = \int_{-\infty}^{\infty} \frac{d\omega}{i\pi}\, \frac{\chi''_{MM}(k\omega)}{(\omega - z)(\omega - i\varepsilon)}\, B^{\text{ext}}(\mathbf{r}). \tag{6}$$

Substituting (4) into (6) and letting $\varepsilon \to 0$ we obtain

$$M(\mathbf{r}z) = \frac{1}{iz}\left[\frac{\chi_{MM}(kz)}{\chi_{MM}(k0)} - 1 \right]\delta\langle M(\mathbf{r}0)\rangle. \tag{7}$$

By the general dispersion relation we proved in the last section, we may write this in the form

$$M(\mathbf{r}z) = \frac{iz}{z^2 - c_0^2(k)\, k^2 + iD(kz)\, k^2 z}\, \delta\langle M(\mathbf{r}0)\rangle \tag{8}$$

where we have called $\gamma(z) \to D(kz)\, k^2$ and $\omega_0{}^2 \to c_0{}^2(k)\, k^2$.

This expression may be readily compared with the expression predicted by hydrodynamics that is, by the ordinarily assumed constitutive equation

$$\mathbf{J}^M(\mathbf{r}t) = -D\nabla M(\mathbf{r}t).$$

If this equation is valid, then the conservation law

$$\frac{\partial}{\partial t}\delta\langle M(\mathbf{r}t)\rangle + \nabla \cdot \langle \mathbf{J}^M(\mathbf{r}t)\rangle = 0 \tag{9}$$

implies the diffusion equation

$$\frac{\partial}{\partial t}\delta\langle M(\mathbf{r}t)\rangle - D\nabla^2\, \delta\langle M(\mathbf{r}t)\rangle = 0. \tag{10}$$

Using the transform defined in (6), we obtain

$$-iz\, M(\mathbf{r}z) - D\nabla^2\, M(\mathbf{r}z) = \delta\langle M(\mathbf{r}0)\rangle \tag{11}$$

or equivalently, since $\delta\langle M(\mathbf{r}0)\rangle$ has only Fourier component k,

$$M(\mathbf{r}z) = \frac{iz}{z^2 + iDk^2z}\, \delta\langle M(\mathbf{r}0)\rangle. \tag{12}$$

Thus if the hydrodynamic law is correct, a measurement of the correlation function or a correct calculation of it for small z and k, will show that $c_0^2(k) = 0$ and determine the limiting value D of $D(kz)$.

Correspondingly, for small k and ω, we obtain

$$\chi''_{MM}(k\omega) = \frac{\chi_{MM}(k0)\,\omega Dk^2}{\omega^2 + (Dk^2)^2} \tag{13}$$

which expresses the behavior of the correlation function if the hydrodynamic law is valid. In this case, we may also express the diffusion constant as

$$\chi_{MM}(00)\,D = \lim_{\omega \to 0}\lim_{k \to 0}\frac{\omega \chi''_{MM}(k\omega)}{k^2} = \lim_{\omega \to 0}\lim_{k \to 0}\frac{\chi''_{JMJM}(k\omega)}{\omega}$$

$$= \lim_{\omega \to 0}\lim_{k \to 0}\frac{\beta \varphi_{JMJM}(k\omega)}{2} = \lim_{\omega \to 0}\lim_{k \to 0}\frac{\beta}{4}\int d\mathbf{r}\,e^{-i\mathbf{k}\cdot\mathbf{r}}\int dt\,e^{i\omega t}\{\hat{\mathbf{k}}\cdot\mathbf{J}^M(\mathbf{r}t),\hat{\mathbf{k}}\cdot\mathbf{J}^M(00)\}. \tag{14}$$

The last expression is known as a Kubo formula; it expresses a transport coefficient in terms of a time integral of the current of the conserved variable. The more physically accessible formula (13) can be studied with neutrons.

Another point bears mention although it is not rigorous even when hydrodynamics is. We have indicated in our oscillator example that a crude but reasonable interpolation scheme is one in which we relate the large and small z behavior by a single collision time parameter. The interpolation scheme has no effect in the hydrodynamic region, but it does give an improved value for $D(kz)$ for large z. It also gives a relation between D, the high frequency behavior, and the relaxation time τ. Specifically, if we write, as in the oscillator,

$$\chi_{MM}(kz) - \chi_{MM}(k0) \cong -\chi_{MM}(k0)\,z^2\left[z^2 + \frac{iz\langle\omega^2_{MM}(k)\rangle\,\tau(k)}{1 - iz\tau(k)}\right]^{-1} \tag{15}$$

we may evaluate $\langle\omega^2_{MM}(k)\rangle \equiv c^2_\infty(k)k^2 = k^2\mu^2 n/m\chi_{MM}(k0)$. In our example, therefore, we have the phenomenological relation

$$D = \frac{\mu^2 n\tau}{m\chi_{MM}} = \frac{c^2_\infty\tau}{\chi_{MM}}. \tag{16}$$

In this formula, and henceforth, we shall occasionally omit $k = 0$ and $\omega = 0$ arguments. Equation (16) can be used to illustrate two points. The first concerns a system like He[3] below its degeneracy temperature and when $\beta\hbar/\tau \ll 1$. Under these conditions the Pauli principle severely restricts scattering and $\tau \sim 1/T^2$. It also reduces the susceptibility in the well known manner from the Curie law $\chi \sim T^{-1}$ to the Pauli result $\chi \sim T^0$. We would consequently expect $D \sim T^{-2}$ and this expectation is borne out quite quantitatively[18].

The second point concerns the behavior of the diffusion constant in the neighborhood of a critical point for a substance that undergoes a ferromagnetic transition. We expect both c_∞^2 and τ, and their combination, the transport coefficient $L = c_\infty^2 \tau$, to be less sensitive than the susceptibility, χ. The diffusion constant is the ratio of this combination which occurs in the phenomenological law $J^M = -L\nabla B^{ext}$, to the susceptibility since $\nabla B^{ext} = (\partial B^{ext}/\partial M) \approx \chi^{-1}\nabla M$. We therefore expect it to vanish. This slowing down is a general phenomenon[19]. It occurs in a critical mixture where $J = -L\nabla\mu$ and $D = L/(\partial c/\partial \mu) \to 0$ because the derivative of the concentration, c, with respect to the chemical potential diverges. It also occurs in thermal conduction at the liquid-gas critical point where $J^\epsilon = -\varkappa \nabla T$ and $D = \varkappa/mnc \to 0$ because c, the specific heat, diverges.

2. Transverse Waves

Let us consider a second example[20], in which the parameter c_0 of Eq. 8 plays a more crucial role, and in which we can make a more definite identification of the external force with a thermodynamic parameter. Suppose we have a fluid in which the momentum density is denoted by $\mathbf{g}(\mathbf{r})$. We may divide $\mathbf{g}(\mathbf{r})$ into two parts, a transverse part $\mathbf{g}^T(\mathbf{r})$ and a longitudinal part $\mathbf{g}^L(\mathbf{r})$, with

$$\nabla \cdot \mathbf{g}^L = \nabla \cdot \mathbf{g}, \qquad \nabla \cdot \mathbf{g}^T = 0,$$

$$\nabla \times \mathbf{g}^T = \nabla \times \mathbf{g}, \qquad \nabla \times \mathbf{g}^L = 0,$$

$$\mathbf{g} = \mathbf{g}^T + \mathbf{g}^L. \tag{17}$$

The momentum conservation law is

$$\frac{\partial g_i(\mathbf{r})}{\partial t} + \nabla_j T_{ij}(\mathbf{r}) = 0 \tag{18}$$

where T_{ij} is the stress tensor. For a homogeneous viscoelastic medium, and for the transverse momentum, the generally assumed phenomenological law at low frequencies and long wavelengths gives

$$\frac{\partial^2 \langle g_i^T(\mathbf{r})\rangle_{n.e.}}{\partial t^2} - \nabla_j \cdot \left\{ \mu \nabla_j \langle v_i^T(\mathbf{r})\rangle_{n.e.} + \eta \nabla_j \frac{\partial \langle v_i^T(\mathbf{r})\rangle_{n.e.}}{\partial t} \right\} = 0 \tag{19}$$

with μ the shear modulus, η the shear viscosity, and v^T the transverse velocity. After an external force which produces the non-equilibrium flow has been turned off,

$$\langle g_i^T(\mathbf{r})\rangle_{n.e.} = mn\langle v_i^T(\mathbf{r})\rangle_{n.e.} \tag{20}$$

and therefore, in analogy with (A6) we have

$$\frac{\partial^2 \langle g_i^T\rangle_{n.e.}}{\partial t^2} - \frac{\mu}{mn}\nabla^2 \langle g_i^T\rangle_{n.e.} - \frac{\eta}{mn}\nabla^2 \frac{\partial \langle g_i^T\rangle_{n.e.}}{\partial t} = 0. \tag{21}$$

Indeed, if the external perturbation is of the form $\int \mathbf{g}(\mathbf{r}t) \cdot \mathbf{v}^{\mathrm{ext}}(\mathbf{r}t)$, and it has not been turned off, the linearized equation reads instead,

$$\frac{\partial^2 \langle g_i^T \rangle_{\mathrm{n.e.}}}{\partial t^2} - \frac{\mu}{mn} \nabla^2 \langle g_i^T \rangle_{\mathrm{n.e.}} - \frac{\eta}{mn} \nabla^2 \frac{\partial \langle g_i^T \rangle_{\mathrm{n.e.}}}{\partial t} = mn \frac{\partial^2 v_i^{\mathrm{ext}T}}{\partial t^2}. \tag{22}$$

We shall not actually derive this equation but merely ask the reader to accept it in view of its electromagnetic analog (i.e. $m \partial^2 v / \partial t^2 = e \partial E^{\mathrm{ext}} / \partial t = -e \partial^2 A^{\mathrm{ext}} / \partial t^2 c$). Also, as we know from the electromagnetic analogy, when v^{ext} is present, the momentum is not described by the operator appropriate when there is no field but by that operator minus $mn \, v^{\mathrm{ext}}$.

We may now proceed in analogy with section A. We may write

$$\frac{\partial^2}{\partial t^2} \tilde{x}(\mathbf{r}\mathbf{r}'; t - t') - \frac{\mu}{mn} \nabla^2 \tilde{x}(\mathbf{r}\mathbf{r}'; t - t') - \frac{\eta}{mn} \nabla^2 \frac{\partial \tilde{x}}{\partial t} (\mathbf{r}\mathbf{r}'; t - t')$$

$$= mn \frac{\partial^2}{\partial t^2} \delta(\mathbf{r} - \mathbf{r}') \, \delta(t - t') \tag{23}$$

since we will then have

$$\langle g_i^T(\mathbf{r}t) \rangle_{\mathrm{n.e.}} = \int \tilde{x}(\mathbf{r}\mathbf{r}'; t - t') \, v_i^{\mathrm{ext}T}(\mathbf{r}'t') \, d\mathbf{r}' dt'. \tag{24}$$

As a result the space-time Fourier transform

$$\tilde{x}(\mathbf{r}\mathbf{r}'; t - t') = \int \frac{d\mathbf{k}}{(2\pi)^3} \int \frac{d\omega}{2\pi} e^{i\mathbf{k}\cdot(\mathbf{r}-\mathbf{r}')-i\omega(t-t')} x(k\omega) \tag{25}$$

satisfies

$$\left(-z^2 + \frac{\mu}{mn} k^2 - \frac{\eta}{mn} izk^2 \right) x(kz) = -mnz^2. \tag{26}$$

If, as always, we denote the retarded commutator of the momentum when $v^{\mathrm{ext}T}$ vanishes by $\tilde{\chi}_{g^T g^T}$, and take account of the additional term in first order perturbation theory noted above which depends explicitly on v^{ext} we may identify

$$-\tilde{\chi}_{g^T g^T} = \tilde{x} - mn \, \delta(\mathbf{r} - \mathbf{r}') \, \delta(t - t') \tag{27}$$

or, introducing the short-hand $\chi_{g^T g^T} \equiv \chi_T$,

$$\chi_T(kz) = -\left[\frac{mnz^2}{z^2 - (\mu - i\eta z) k^2/mn} - mn \right] = \frac{(\mu - i\eta z)k^2}{-z^2 + (\mu - i\eta z) k^2/mn}$$

$$\frac{\chi_T''(k\omega)}{\omega} = \frac{\eta k^2 \omega^2}{(\omega^2 - \mu k^2/mn)^2 + (\eta k^2 \omega/mn)^2}. \tag{28}$$

In the long wavelength limit, we can deduce from Eq. (28) the two additional identities specific to conserved densities:

(A) a reversible thermodynamic identity or sum rule (a fluctuation formula)

$$\lim_{k\to 0}\lim_{\omega\to 0}\chi_T(k\omega) = \lim_{k\to 0}\int \frac{\chi_T''(k\omega)}{\omega}\frac{d\omega}{\pi} = \left(\frac{\partial g^T}{\partial v^T}\right) = mn. \tag{29}$$

(B) An irreversible thermodynamic identity (a Kubo formula)

$$\lim_{\omega\to 0}\lim_{k\to 0} -\frac{i\omega}{k^2}\chi_T(k\omega) = \frac{i\mu}{\omega} + \eta = \lim_{\omega\to 0}\lim_{k\to 0}\int \frac{d\omega'}{\pi i}\frac{\omega\chi_T''(k\omega')}{(\omega'-\omega)k^2}. \tag{30}$$

We may write (A) alternatively in the form

$$\lim_{k\to 0}\chi_T''(k\omega)/\omega = mn\delta(\omega)\,\pi = \left(\frac{\partial g^T}{\partial v^T}\right)\pi\,\delta(\omega) \tag{31}$$

since as $k \to 0$, $\chi_T''(k\omega)$ vanishes unless $\omega = 0$ because g is a conserved quantity. (The fastidious reader might note that the analog to (29) in the magnetic example is not correct. The quantity $\chi_{MM}(k0)$ gives the ratio of the magnetization to the external field. The long range magnetic interaction makes it necessary to distinguish between the external and total fields. We shall have more to say about this problem when we discuss electromagnetic properties.) We may write (B) alternatively as

$$\lim_{\omega\to 0}\lim_{k\to 0}\frac{\omega}{k^2}\chi_T''(k\omega) = \eta;\ \lim_{\omega\to 0}\lim_{k\to 0} -\frac{\omega^2}{k^2}\chi_T'(k\omega) = \mu \tag{32}$$

or as, for small ω,

$$\lim_{k\to 0}\frac{\omega}{k^2}\chi_T''(k\omega) = \eta + \mu\delta(\omega)\pi. \tag{33}$$

There are many other ways to write the identity (B). For example, since

$$\lim_{\omega\to 0}\lim_{k\to 0}\omega\chi_T''(k\omega)/k^2 = \lim_{\omega\to 0}\lim_{k\to 0}\tfrac{1}{2}\beta\omega^2\varphi_T(k\omega)/k^2 \tag{34}$$

we may write

$$\eta = \lim_{\omega\to 0}\lim_{k\to 0}\frac{\beta}{4}\int d\mathbf{r}\int_{-\infty}^{\infty} dt\, e^{-i\mathbf{k}\cdot\mathbf{r}+i\omega t}\frac{1}{2}(\delta_{jm} - \hat{k}_j\hat{k}_m)$$

$$\times \langle\{\hat{k}_i T_{ij}(\mathbf{r}t) - \hat{k}_i\hat{k}_l\hat{k}_j T_{il}(\mathbf{r}t),\ \hat{k}_m T_{mn}(00) - \hat{k}_m\hat{k}_n\hat{k}_p T_{np}(00)\}\rangle. \tag{35}$$

Since $\lim_{k\to 0}\omega\chi_T''/k^2 = \eta + \pi\mu\delta(\omega)$ for small ω we can also write

$$\eta = \int_{-\infty}^{\infty} dt\left[\lim_{k\to 0}\frac{\beta}{4}\int d\mathbf{r}\, e^{-i\mathbf{k}\cdot\mathbf{r}}\frac{1}{2}(\delta_{jm} - \hat{k}_j\hat{k}_m)\langle\{\ \}\rangle - \frac{\mu}{2}\right]. \tag{36}$$

Just as this last expression takes account of the fact that the value of $\omega\chi''/k^2$ for $\omega = 0$ is different from its limit as $\omega \to 0$, it is frequently necessary to distinguish between the value for $k = 0$ and the limit as $k \to 0$. The value for $k = 0$ often depends on the boundaries and thus on such unphysical questions as whether the correlations are evaluated in a canonical or grand canonical ensemble. Such considerations are clearly of no physical interest and do not affect the limiting values. Since we are discussing physics we will always be concerned with the latter.

We have chosen the transverse momentum for our second illustration because it is uncoupled in the hydrodynamic limit from the other conserved quantities in a fluid—the energy, mass, and longitudinal momentum. The fact that its current, the part of the stress tensor whose divergence equals the rate of change of transverse momentum, involves several indices is an unfortunate and inessential aspect of Eqs. (35) and (36). We trust that these indices which identify the current for the transverse momentum will not discourage the reader.

We wish to stress that these correlation functions of conserved quantities also have all the properties we derived for an arbitrary correlation function. Indeed Eq. (28) is similar to the general dispersion relation (B. 46)

$$-\chi_T(kz) + \chi_T(k0) = \frac{z^2\chi_T(k0)}{z^2 - c_{T0}^2(k)k^2 + izD_T(kz)k^2} \tag{37}$$

in which we have made the replacements

$$D_T(kz)k^2 = \gamma(kz) \quad c_{T0}^2(k)k^2 = \omega_0^2 \tag{38}$$

to stress the fact that as $k \to 0$, χ must vanish except at $z = 0$.

The quantity $D(kz)$ plays the role of a complex frequency and wave number dependent diffusion constant. Indeed, it seems reasonable in the case of the transverse momentum to call the combination corresponding to $\gamma(z)m$, a complex shear viscosity, i.e., to call

$$\chi_T(k0)D_T(kz) = \eta(kz). \tag{39}$$

As z approaches the real axis we may write

$$\chi_T(k0) [D_T'(k\omega) + iD_T''(k\omega)] = \eta'(k\omega) + i\eta''(k\omega). \tag{40}$$

The combination corresponding to $\Gamma(z)m$ is then

$$\chi_T(k0) [D_T(kz) + ic_{T0}^2(k)/z] = \eta(kz) + i\chi_T(k0)c_{T0}^2(k)/z$$
$$\to \eta'(k\omega) + i\eta''(k\omega) + i\chi_T(k0)c_{T0}^2(k)/\omega + \pi\delta(\omega)\chi_T(k0)c_{T0}^2(k). \tag{41}$$

It is necessary to separate out the term $i c_{T0}^2(k)/z$, only when the substance supports shear waves in the hydrodynamic regime; it is therefore unnecessary in simple fluids. Corresponding to the sum rules for $\gamma'(\omega)$ we have

$$k^2 \int \frac{d\omega}{\pi} D_T'(k\omega) = \frac{1}{\chi_T(k0)} \int \frac{d\omega}{\pi} \omega \chi_T''(k\omega) - c_{T\infty}^2(k)k^2.$$

$$\equiv k^2 [c_{T\infty}^2(k) - c_{T0}^2(k)] \tag{42}$$

$$k^2 \int \frac{d\omega}{\pi} \omega^2 D_T'(k\omega) = \frac{1}{\chi_T(k0)} \int \frac{d\omega}{\pi} \omega^3 \chi_T''(k\omega) - [c_{T\infty}^2(k)k^2]^2. \tag{43}$$

We also have the interpolation formula, in terms of a single collision time $\tau(k)$. For Im $z > 0$, this formula is

$$D_T(kz) = \frac{[c_{T\infty}^2(k) - c_{T0}^2(k)] \tau(k)}{1 + iz\tau(k)} \; ; \; D_T'(k\omega) = \frac{[c_{T\infty}^2(k) - c_{T0}^2(k)] \tau(k)}{1 + \omega^2\tau^2(k)}$$

$$\tag{44}$$

$$-\frac{\chi_T(kz)}{\chi_T(k0)} + 1 = \frac{z^2}{z^2 - c_{T0}^2(k)k^2 + izk^2 D_T(kz)} \tag{45}$$

$$\frac{\chi_T''(k\omega)}{\chi_T(k0)} = \frac{\omega^3 k^2 [1 + (\omega\tau^2)] [c_{T\infty}^2 - c_{T0}^2]}{[(\omega^2 - c_{T0}^2 k^2) + (\omega\tau)^2 (\omega^2 - c_{T\infty}^2 k^2)]^2 + [(c_{T\infty}^2 - c_{T0}^2)k^2\omega\tau]^2}$$

$$\tag{46}$$

with

$$\eta = mn \, D(00); \quad \mu = mn \, c_T^2(0); \quad mn = \chi_T(00)$$

in the hydrodynamic limit. Plotted in Fig. 15 are $\chi_T''(k\omega)$ for small k and ω in a simple gas, and in an isotropic solid which supports hydrodynamic shear waves.

The quantity $c_{T\infty}^2(k)$, like ω_∞^2 can be calculated directly for a system interacting by two-body forces. Likewise, the quantity $\chi_T(k0)$ is a thermodynamic quantity. Classically, the latter is always mn, expressing the classi-

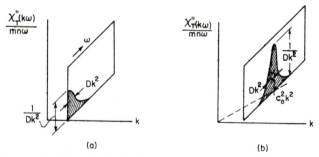

(a)   (b)

Fig. 15. Behavior of $\chi_T''(k\omega)$ in the hydrodynamic regime for (a) a weakly interacting gas or a liquid, and (b) a viscoelastic, isotropic medium which supports shear waves. At long wavelengths the modes exhaust the rule sum and their shape is Lorentzian.

cal independence of momentum and position. The former[22] is similar to $\langle \omega_{vv}^2 \rangle$ (discussed after (B. 55) and given by (A. 33))

$$mc_{T\infty}^2(k) = \frac{2}{3} \frac{\langle KE \rangle}{\langle N \rangle} + n \int d\mathbf{r}\, g^{(2)}(r) \frac{\sin^2 (\tfrac{1}{2}\mathbf{k} \cdot \mathbf{r})}{k^2} [\nabla^2 - (\hat{\mathbf{k}} \cdot \nabla)^2]\, v(r).$$

(47)

The analog to the formula for $\langle \omega_{vv}^4 \rangle$ is also given by a lengthy formula. Nevertheless this moment has also been evaluated[23], and employed in a calculation of the viscosity with the same exponential relaxation time approximation discussed in Sec. B for the friction constant. At liquid densities ($\varrho \sim 1$ gm/cc), the calculated shear viscosity of argon is also about 10–20 percent from the experimental value over a range of temperatures. We refer the interested student to the literature[24].

In the remainder of this section, we shall consider the non-hydrodynamic ($kl \gg 1$ and $\omega\tau \gg 1$) behavior of $\chi_T''(k\omega)/\omega$ more accurately for both the gas and solid. Our interpolation formula is not particularly justified in this limit. In the gas, for example, we will have essentially free particle behavior with

$$\chi_T''(k\omega) = mn \left(\frac{\omega}{kv} \right) \left[\frac{\pi}{2} \right]^{\frac{1}{2}} \exp\left[-\frac{1}{2} \left(\frac{\omega}{kv} \right)^2 \right]$$

(48)

[$v = (\beta m)^{-\frac{1}{2}}$ is the thermal velocity]. In other words for $kl > 1$ or $\omega\tau > 1$ where l is a mean free path ($l \sim v\tau$) the function $\chi_T''(k\omega)$ will be Gaussian rather than Lorentzian and its frequency width will be determined primarily by the thermal velocity. This is only reasonable since until the first collision the particles are spreading apart with thermal velocities in a gas. It is only after many collisions that the transverse momentum satisfies a hydrodynamic, diffusion equation.

In the isotropic solid the distinction between $\omega\tau \ll 1$ and $\omega\tau \gg 1$ is less pronounced. There is always phonon like behavior, and in the neighborhood of the phonon peak $\chi_T''(k\omega)$ is Lorentzian. However, while the Lorentzian peak contains all of the strength in the hydrodynamic limit it contains only some of it when $\omega\tau \gg 1$. Moreover the width of the peak behaves quite differently. For example[25] at low temperatures, instead of behaving in the hydrodynamic fashion as $\eta k^2/mn$, it behaves linearly with k (when $\beta\hbar ck \ll 1$) and also as β^{-4}. If we could measure $\chi_T''(k\omega)$ everywhere we could see the transition between these regimes and when $kl \gg 1$, we could determine the strength of the resonance (the renormalization constant $Z(k)$), as well as the attenuation $\bar{\gamma}'(k)$. We could also examine the behavior of the spectral function away from the resonance and check the calculations which have been done for this behavior. This is probably impossible. What has been done, however is to measure the temperature and frequency dependence of the width of the resonant peak when $\omega\tau \gg 1$,

7

that is, the attenuation of ultrasonic shear waves. This exhibits the predicted non-hydrodynamic behavior, which is plotted schematically in Fig. 16.

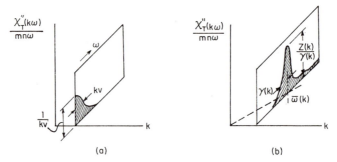

(a)   (b)

Fig. 16. Behavior of $\chi_T''(k\omega)$ in the non-hydrodynamic regime for the same systems shown in Fig. 15. (a) For the weakly interacting gas the behavior is like that of a free gas: Gaussian, with a width proportional to the thermal velocity v and the wavenumber k. In terms of a mean free time τ the hydrodynamic width $Dk^2 \sim v(v\tau k) \, k$ overestimates the result in the non-hydrodynamic ($kv\tau \gg 1$) regime. A parametric Lorentzian fit to $\chi_T''(k\omega)$ of the form discussed in the text therefore requires $\tau^{-1}(k) \approx \tau^{-1}(0) + kv$. (b) For the iso-tropic solid, phonons persist in the non-hydrodynamic regime but their damping is also smaller than hydrodynamics predicts; the strength of the phonon peak may be substantially reduced by anharmonic effects with "many-phonon" contributions giving the remainder of the sum rule.

D. Correlations of Conserved Quantities, Particularly the Density

The examples we quoted in the previous section are discussed in a paper by Kadanoff and myself[26]. Also discussed in that paper are the corresponding formulas for the remaining hydrodynamic parameters — the energy and density. Of these correlation functions, the one that is most easily accessible to measurement is the density correlation function. In these lectures we will not have time to go through the hydrodynamic analysis which leads to predictions for the density correlation function like those obtained for χ_T and χ_{MM} in Section C. Instead we shall content ourselves with quoting the results for χ_{nn} and commenting on their experimental significance,

Before doing so, however, let us merely state the generalization its derivation entails. The chief new feature is the necessity for treating simultaneously the conservation laws for all densities whose currents depend phenomenologically on common density gradients, and which therefore lead to coupled linearized hydrodynamic equations. Because of this coupling, the correlation functions of a particular density, like those for a particular coupled oscillator in Section A, exhibit several normal modes with varying strength.

In the thermodynamic discussion of a fluid we would encounter difficulties if we tried to describe a two component system using only the laws of mass and energy conservation and the associated thermodynamic description, or if we tried to describe a one component fluid as we describe a photon gas, using only the law of energy conservation and not specifying the mass density. Likewise, in a ferromagnet, it would be inadequate to employ an ensemble in which the magnetization direction is unspecified. For each system an ensemble must be employed which stipulates all conserved quantities. Only when they are all taken into account, and experimentally controlled, will measurements give well defined results with small fluctuations and correlations have finite range. Correlations in a ferromagnet extend over large distances; it is only when the direction of magnetization is stipulated that the remaining correlations have finite range, and experimental results microscopic significance.

When the ensemble is described by a certain number of conservation laws, say n, the number of thermodynamic correlation functions is of order n^2. Correspondingly, there are about n^2 thermodynamic second derivatives (like compressibilities and susceptibilities) and a similar number of irreversible or Onsager coefficients. The number of independent terms is actually somewhat smaller because of a number of symmetry properties[27]. A general discussion of thermodynamics, both reversible and irreversible must take account of these thermodynamic cross derivatives and terms like thermal diffusion coefficients which relate currents of one conserved quantity to derivatives of another. It must also be concerned with the effect on the frequency and damping of the n "hydrodynamic normal modes" like sound propagation in which oscillations of the various conserved quantities participate. For example, it is the coupling of density and energy fluctuations which leads to the replacement of the Newton sound velocity $c^2 = (dp/dmn)_T$ by the Laplace sound velocity $c^2 = (dp/dmn)_s$. Likewise, it is this coupling which leads to a temperature diffusivity $D = \varkappa/mnc_p$ in place of $\varkappa/mnc_v$, and to an attenuation of longitudinal sound by thermal conduction, as well as by longitudinal viscosity. Specifically, the frequencies of the two coupled longitudinal normal modes in a fluid are given approximately by

$$\omega^2 - c_i^2 k^2 + iD_i k^2 \omega = 0 \qquad (1)$$

where, in terms of the bulk viscosity ζ, the shear viscosity η, the thermal conductivity $\varkappa$, the specific heats at constant pressure c_p, and volume, c_v, we have

$$mc_1^2 = \left(\frac{dp}{dn}\right)_s ; \qquad D_1 = \frac{\zeta + (4\eta/3)}{mn} + \frac{\varkappa}{mn}\left(\frac{1}{c_v} - \frac{1}{c_p}\right), \qquad (2)$$

and

$$c_2^2 = 0; \qquad D_2 = \frac{\varkappa}{mnc_p}. \qquad (3)$$

The quantity s is the entropy per unit mass.

7*

[In superfluids, apart from dissipative terms the corresponding equations[28] are

$$m(c_1^2 + c_2^2) = \frac{mTn_s}{n_n} \frac{s^2}{c_v} + \left(\frac{dp}{dn}\right)_s$$

$$mc_1^2 c_2^2 = \frac{Tn_s}{n_n} \frac{s^2}{c_v} \left(\frac{dp}{dn}\right)_T$$

where n_s is the superfluid density and $n_s + n_n = n$.]

The comparison of hydrodynamics and correlation functions to which we alluded above then gives

$$\chi_{nn}''(k\omega) = n\left(\frac{\partial n}{\partial p}\right)_T \left[\frac{D_2 k^2 \omega [1 - (c_v/c_p)]}{\omega^2 + (D_2 k^2)^2} + \frac{D_1 k^4 \omega c_1^2 (c_v/c_p)}{(\omega^2 - c_1^2 k^2)^2 + (D_1 k^2 \omega)^2}\right]$$

$$- n\left(\frac{\partial n}{\partial p}\right)_T \frac{D_2 k^2 \omega(\omega^2 - c_1^2 k^2)[1 - (c_v/c_p)]}{(\omega^2 - c_1^2 k^2)^2 + (D_1 k^2 \omega)^2}, \tag{4}$$

$$\chi_{n\varepsilon}''(k\omega) = T\left(\frac{\partial n}{\partial T}\right)_p \left[\frac{D_2 k^2 \omega}{\omega^2 + (D_2 k^2)^2} - \frac{D_1 k^2 \omega(\omega^2 - c_1^2 k^2)}{(\omega^2 - c_1^2 k^2)^2 + (D_1 k^2 \omega)^2}\right]$$

$$+ \frac{\varepsilon + p}{n} \chi_{nn}''(k\omega), \tag{5}$$

and

$$\chi_{\varepsilon\varepsilon}''(k\omega) = \frac{mnc_p T D_2 k^2 \omega}{\omega^2 + (D_2 k^2)^2} + 2\frac{\varepsilon + p}{n} \chi_{n\varepsilon}''(k\omega) + \left(\frac{\varepsilon + p}{n}\right)^2 \chi_{nn}''(k\omega) \tag{6}$$

where ε is the energy density.

In the low wave number-low frequency limit, the correlation function composed of the transverse component of the momentum exhibits a diffusion structure with diffusivity, $D_T = \eta/mn$, given by the viscosity divided by the mass density. The correlation functions above also have a diffusion structure but here the diffusivity is the thermal diffusivity, $D_2 = \varkappa/mnc_p$. They also exhibit the damped sound wave propagation. The total weight of χ_{nn}''/ω is $n(\partial n/\partial p)_T$ of which a proportion $(1 - c_v/c_p)$ comes from the diffusion process and a proportion c_v/c_p comes from the sound propagation.

Note once more that the hydrodynamic analysis is only correct in the limit as $k \to 0$. Thus, for example χ_T behaves asymptotically in ω as $i\eta k^2/\omega$ (which vanishes as $k \to 0$), while rigorously χ_T behaves like ω^{-2} for all k.

Eq. (4) is an old and famous result derived in 1934 by Landau and Placzek[29] and depicted in Fig. 17. Indeed, using equation (4) we can determine by measuring $\chi_{nn}''(k\omega)/\omega$, for small k and small ω,

$$n\left(\frac{\partial n}{\partial p}\right)_T, \quad \frac{c_p}{c_v}, \quad \frac{\varkappa}{mnc_p}, \quad \frac{\zeta + (4\eta/3)}{mn}. \tag{7}$$

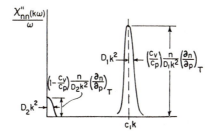

Fig. 17. The characteristic Landau-Placzek expression for $\chi_{nn}''(k\omega)/\omega$ involves a Brillouin doublet ($\omega = \pm c_1 k$) and a central peak. The widths give the damping of these modes; the total intensity is given by $n(\partial n/\partial p)_T$ and is exhausted by the two peaks whose relative intensities are (c_v/c_p) and $[1 - (c_v/c_p)]$ respectively. The central peak would be replaced by a central doublet in a superfluid.

In addition since $\pi^{-1} \int d\omega\, \omega \chi_{nn}''(k\omega) = mnk^2$ we could determine mn. If we also measure $\chi_{nn}''(k\omega)/\omega$ for large k, so that we have the instantaneous correlation function

$$\tilde{S}_{nn}(k, t = 0) = \int \frac{2\hbar\omega}{1 - e^{-\beta\hbar\omega}} \frac{\chi_{nn}''(k\omega)}{\omega} \frac{d\omega}{\pi} \qquad (8)$$

we can also determine in a classical system with a known interaction potential, the energy and pressure, and therefore the specific heat. The quantity $(\partial n/\partial T)_p$ can be determined by measuring χ_{nn}'' as a function of T, and the only remaining parameter, η, can be determined, as we saw in Sec. C from $\chi_T''(k\omega)$.

To state this result more theoretically: all thermodynamic and hydrodynamic parameters of a classical fluid as well as most other measurable properties can be determined by measuring (experimentally) or calculating (theoretically) the function $\chi_{g_i g_j}''(k\omega)$ or $S_{g_i g_j}(k\omega)$. (In a quantum fluid, a measurement of the specific heat and pressure would also be necessary, since the kinetic energy density and kinetic pressure are not just $\frac{3}{2}nkT$ and nkT.)

The function $\chi_{nn}''(k\omega)/\omega$ is an even function. Therefore $S_{nn}(k\omega)$ is approximately even (i.e. it is even when $\hbar\omega \ll kT$). The function $\chi_{nn}''(\omega)/\omega$ therefore has two peaks at $\omega = \pm c_1 k$ (a Brillouin doublet) and a central peak.

Also, since

$$\frac{c_v}{c_p} = \left[1 + \frac{T}{mn^2 c_v}\left(\frac{dp}{dT}\right)_n^2 \left(\frac{\partial n}{\partial p}\right)_T\right]^{-1} \qquad (9)$$

at low temperatures, the central peak is vanishingly small. Typically it behaves at $(T/T_{\text{Debye}})^4$ for small T. At $T = 0$, the doublet in χ'' at $\omega = \pm ck$ (with $(dp/dn)_T = mc^2$ and $\omega \chi_{nn}''(k\omega) = nk^2 \pi|\omega|\delta(\omega^2 - c^2k^2)/m$) exhausts in the long wavelength limit the sum rules

$$\lim_{k \to 0} \int \frac{\chi_{nn}''(k\omega)}{\omega} \frac{d\omega}{\pi} = n\left(\frac{dn}{dp}\right)_T$$

and

$$\int \chi_{nn}''(k\omega)\omega \frac{d\omega}{\pi} = \frac{nk^2}{m}. \qquad (10)$$

The opposite extreme occurs at the critical point of a fluid. Then $(\partial n/\partial p)_T = \infty$ so that the area under $\chi_{nn}''(k\omega)$ becomes infinite. The great preponderance of this area comes from the central peak, the area under the Brillouin peaks remaining relatively constant. Although the area under the central peak increases and its tail tends to swamp the Brillouin peaks its half width, which depends on $\varkappa/mnc_p$, is reduced because c_p is increased. The behavior of the correlation function in this region has been the subject of intensive study recently[30]. Likewise the behavior in critical mixtures, which also show increased scattering and slowed diffusion, has been recently investigated. (See Fig. 18.)

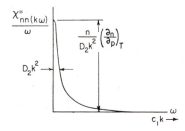

Fig. 18. The same Landau-Placzek formula when $n(\partial n/\partial p)_T \to \infty$ near a critical point and $D_2 \to 0$.

Implicit in the above discussion has been the assumption that the behavior near T_c is dominated by the zero in $(\partial p/\partial n)_T$. While this assumption is born out, $(\partial p/\partial n)_T \sim (T - T_c)^\gamma$ where $\gamma \sim 1.3$; it is not unity as it would be in simple theories. Likewise, it appears that while $\varkappa$ and c_v are not regular at T_c, they do not diverge very strongly. The experimental evidence seems to indicate that $c_v \sim (T - T_c)^{-\alpha}$ where $\alpha \sim .1$, $\varkappa \sim (T - T_c)^{-\lambda}$ where λ lies between .1 and .7 but the results are not definitive. The weak singularities may also affect the position of the Brillouin peaks near T_c, (the velocity of sound), but the attenuation, and the dominance of the central peak makes this difficult to discuss.

Like the function $\chi_T(kz)$ we discussed earlier, the function χ_{nn} and the related function $\chi_L(kz)$

$$m^2z^2\,\chi_{nn}(kz) = k^2\chi_L(kz) - nmk^2 \qquad (11)$$

may be studied outside of the hydrodynamic regime, that is, for $\omega\tau \gg 1$ and $kl \gg 1$. In a rare gas, the transition occurs for relatively small k. For larger k the behavior is again free-gas like. In particular

$$\frac{\chi_{nn}''(k\omega)}{\omega} = \left[\frac{\pi}{2}\right]^{\frac{1}{2}} \frac{n\beta}{kv} \exp\left[-\frac{1}{2}\left(\frac{\omega}{kv}\right)^2\right] \qquad (12)$$

For the rare gas it is possible to interpolate between these limits using the Boltzmann equation with different force laws, and various other approximations. In Figs. 19–21, are plotted theoretical curves showing how the transition takes place.[32] Also plotted for comparison are some experimental studies in the Brillouin region and the transition region.[33]

In an isotropic solid one can also study $\chi_L''(k\omega)/\omega$ in the nonhydrodynamic regime, and, at sufficiently high frequencies, one finds behavior of the same form indicated for $\chi_T''(k\omega)/\omega$. Actually the situation is considerably more complicated; there are various regimes[34] depending on the curvature of $c_\infty^2(k)$ with k^2 as well as on the parameters $\omega\tau$ and $\hbar\omega/kT$.

There is also a particularly interesting domain in the isotropic solid when the temperature is low so that the phonon picture is approximately valid and the non-momentum conserving umklapp processes unimportant. Under these circumstances, a kind of hydrodynamic picture is applicable for $\omega\tau_u \gg 1$, in which the energy current (which is essentially, the momentum density times c^2) is conserved. One then has, when $\omega\tau \ll 1$, essentially a gas of phonons in the hydrodynamic limit.[35] For this gas of phonons the pressure is $\frac{1}{3}$ the energy density so that we may write

$$\frac{\partial\varepsilon}{\partial t} = -\nabla \cdot \mathbf{j}^\varepsilon = -\nabla \cdot c^2 \mathbf{g} = -\nabla^2 \frac{c^2}{3}\varepsilon. \qquad (13)$$

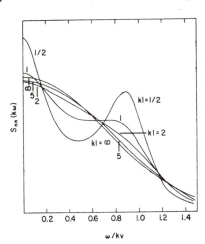

Fig. 19. $S_{nn}(k\omega)$ for various values of wave number k times mean free path, l.

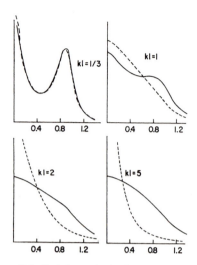

Fig. 20. Deviation of $S_{nn}(k\omega)$ from the hydrodynamic prediction for $kl > 1$.

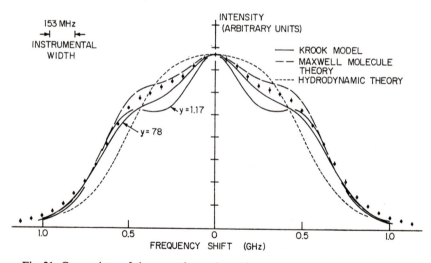

Fig. 21. Comparison of theory and experiment in a rare gas when $kv\tau$ or kl is about unity. The experimental data for xenon are indicated by the dots. The theoretical curves for several models are shown. The Krook model does not give the correct ratio for the viscosity and thermal conductivity; the curve is plotted for two values of the mean free path, (which is essentially y) which fit the two transport coefficients. The Boltzmann equation also gives results depending essentially on a single parameter which is consistently determined by the two transport coefficients. With a Maxwell force law, the depicted curve is found; with a hard core interaction an even better fit has recently been obtained.

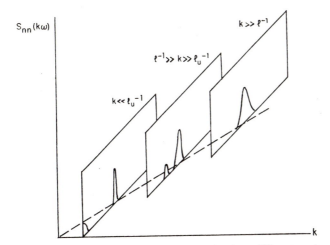

Fig. 22. The schematic dynamic structure function in three different regimes in a low temperature solid. For $k \ll l_u^{-1}$ there is characteristic hydrodynamic behaviour. For larger k, but kl still less than one, many collisions take place but, since $kl_u \gg 1$, none of these alter the energy current. There is then second sound. For still larger k, there are non-hydrodynamic phonons or sound waves, but no second sound. There are also many phonon contributions as in the transverse case.

It follows that the energy density no longer satisfies a diffusion equation but a wave equation. One then has a function $\chi''(k\omega)/\omega$ which takes on the various aspects depicted in Fig. 22. Recent sound propagation experiments[36] in solid helium have demonstrated the existence of the intermediate domain in which second sound occurs. In superfluid helium at low temperatures we have qualitatively the same picture with $\tau_u \to \infty$ so that only the last two domains occur[37].

In liquids, the situation is probably somewhat intermediate. In normal liquid helium 4, there is evidence that the behavior of $S_{nn}(k\omega)$ is quite solid-like.[38] In liquid argon the picture is somewhat more ambiguous but there appear to be indications that $S_{nn}(k\omega)$ may be peaked away from $\omega = 0$, at

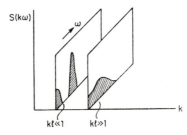

Fig. 23. Schematic picture of variation of "phonons" in a dense liquid. The peak, which appears to vary as shown in Fig. 24, is considerably broadened at wavelengths measured by neutrons and corresponding to a few interparticle spacings.

a position varying with k roughly as the peak in helium varies. This behavior and some experimental results[39,40] concerning it are depicted in Figs. 23 and 24. As is clear from the pictures, in these other fluids, the breadth of the "phonons" is considerably greater.

There is one final formal point on which we should remark, and that is the manner in which the mode coupling we have discussed, which gives rise to pairs of peaks in χ_{nn}, manifests itself in the general dispersion relation which played so central a role in our discussion of χ_{vv} for the oscillator and in χ_T for the transverse momentum. By the same arguments used there we may rigorously conclude that for arbitrary frequency and wavelength

$$1 - \frac{\chi_L(k\omega)}{mn} = \omega^2 \left[\omega^2 - c_{L0}^2(k)k^2 + k^2\omega^2 \int \frac{d\omega'}{\pi} \frac{D_L'(k\omega')}{\omega'^2 - \omega^2} \right]^{-1}$$

$$\frac{\chi_L''(k\omega)}{m^2} = \frac{\chi_{nn}''(k\omega)\omega^2}{k^2} ; \quad \chi_L(k0) = mn \qquad (14)$$

$$\frac{\chi_L''(k\omega)}{mn} = \frac{D_L'(k\omega)k^2\omega^3}{\left[\omega^2 - c_{L0}^2(k)k^2 + k^2\omega^2 P\int \frac{d\omega'}{\pi} \frac{D_L'(k\omega')}{\omega'^2 - \omega^2} \right]^2 + [D_L'(k\omega)k^2\omega]^2} .$$

$$(15)$$

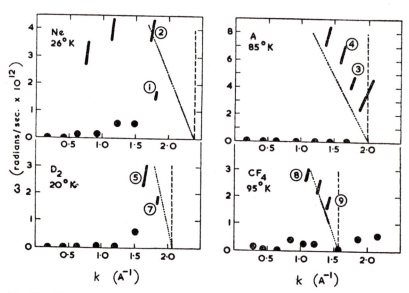

Fig. 24a. The peak values of $S_{nn}(k\omega)$ in various liquids as measured by Chen, *et al.* The dotted line is drawn at the slope determined by the sound velocity. The intercept, indicated by the dashed line coincides with the first maximum in the structure function. See footnote 42.

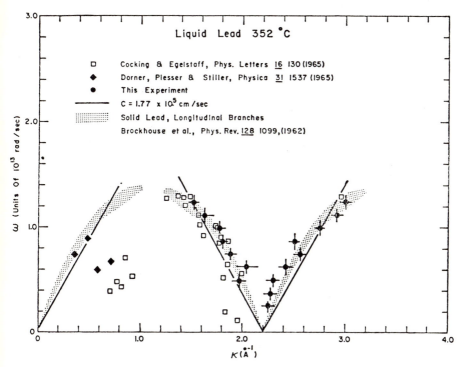

Fig. 24b. The dispersion curve of Randolph and Singwi for liquid lead and a comparison with previous measurements.

We have already shown that[41]

$$\zeta + \frac{4}{3}\eta = \lim_{\omega \to 0} \lim_{k \to 0} [mn \, D_L'(k\omega)] \tag{16}$$

and we can also evaluate for particles interacting through a two-body potential[22]

$$\int \frac{d\omega}{\pi} \, \omega \chi_L''(k\omega)/k^2 = mn c_{L\infty}^2(k)$$

$$= n\left[\frac{2\langle KE \rangle}{\langle N \rangle} + \frac{\hbar^2 k^2}{4m} + n \int d\mathbf{r} \, g^{(2)}(r) \frac{2\sin^2(\frac{1}{2}\mathbf{k}\cdot\mathbf{r})}{k^2} (\hat{\mathbf{k}} \cdot \nabla)(\hat{\mathbf{k}} \cdot \nabla) \, v(r) \right]. \tag{17}$$

Moreover it is easy to prove that[42]

$$c_{LO}^2(k) = \chi_{nn}^{-1}(k0)(n/m).$$

When the wave-like mode is not the only mode of the system, the function D_L' will be more complicated than in our earlier examples. In particular there will be a collision time of the order of k^2 so that although $c_{Lo}^2(k)$ $= (dp/dmn)_T$, the sound velocity at long wavelengths has the value $(dp/dmn)_s$. In a normal fluid the simplest interpolation formula consistent with the hydrodynamical description has a prescribed long wavelength behavior,

$$D_L'(k\omega) = \frac{(\varkappa k^2/mnc_v)\,[(c_p/c_v) - 1]\,c_{LO}^2}{\omega^2 + (\varkappa k^2/mnc_v)^2} + \frac{\zeta + (4\eta/3)}{mn[1 + (\omega\tau)^2]}$$

$$mc_{LO}^2 = \left(\frac{dp}{dn}\right)_T$$

which requires us to identify the viscosity with an expression involving $c_{L\infty}^2$ and τ, and results in

$$\omega^2 = c_{LO}^2 \left(\frac{c_p}{c_v}\right) k^2 = \left(\frac{dp}{dn}\right)_s \frac{k^2}{m}$$

at low wave numbers and frequencies.

When two wave-like modes are coupled, (as in a superfluid or our phonon gas example), the rapidly varying parts of $D_L(k\omega)$ cannot be represented even by a distribution of collision times. In that case $D_L(k\omega)$, like $\chi''(k\omega)$, contains resonant terms. The great freedom in $\chi''(k\omega)/\chi(k0)$ we have discussed is not fully realized by a form which allows only for relaxation times (poles on the negative imaginary axis in the continuation of the smoothed function from the upper half plane into the lower half plane).

For example, in superfluids, at low frequencies and wavelengths, we have[43]

$$D_L'(k\omega) = \frac{(\varkappa k^2/mnc_v)[c_p/c_v) - 1]\,\omega^2 c_{LO}^2}{[\omega^2 - k^2(n_s/n_n)\,(Ts^2/c_v)]^2 + [(\varkappa/mnc_v)k^2\omega]^2} + \frac{\zeta_2 + (4\eta/3)}{mn[1 + (\omega\tau)^2]}.$$

$$(18)$$

In other words such a form requires, as did the coupled oscillators in Sec. A

$$\frac{D_L'(k\omega)}{c_{L\infty}^2(k) - c_{LO}^2(k)} = \sum_i b_i(k)\,\tau_i(k) \frac{\omega^2}{[\omega^2 - \omega_i^2(k)]^2\tau_i^2(k) + \omega^2}. \qquad (19)$$

Thus we see that (a) coupling of modes will lead to a rapidly varying function $D(k\omega)$, and consequently to the fact that $\omega^2 = c_0^2 k^2$ does not describe the pole of χ even at low frequencies, and (b) coupling of wave-like modes leads to resonant behavior in $D(k\omega)$.

These features do not affect the identification of the Onsager coefficient L and its reactive counterpart R (η and μ in the shear example) or the thermo- dynamic derivative $\partial A_j/\partial a_j$($mn$ in the shear example) with the limiting values of χ indicated there. For example, although in $\chi_L(k\omega)$ which equals

$$\frac{k^2\left[-n\left(\frac{dp}{dn}\right)_s \omega + i\left(\zeta + \frac{4}{3}\eta\right)\omega^2 - i\left(\frac{dp}{dn}\right)_s \frac{\varkappa}{c_p}\frac{k^2}{m}\right]}{\left[\omega + \frac{i}{mn}\frac{\varkappa}{c_p}k^2\right]\left[\omega^2 - \left(\frac{dp}{dn}\right)_s \frac{k^2}{m} + \frac{ik^2\omega}{mn}\left\{\frac{4}{3}\eta + \zeta + \varkappa\left(\frac{1}{c_v} - \frac{1}{c_p}\right)\right\}\right]}, \tag{20}$$

the velocity of the sound wave mode is given by $mc^2 = (dp/dn)_s$ and not $mc_{L0}^2 = (dp/dn)_T$, in the limit in which $\omega \to 0$ and then $k \to 0$ we have

$$\lim_{k\to 0} \chi_L(k0) = mn = \frac{\partial g}{\partial v}, \tag{21}$$

$$\lim_{\omega\to 0} \lim_{k\to 0} -(\omega^2/k^2)\, \chi_L(k\omega) = n(dp/dn)_s = R_L, \tag{22}$$

and

$$\lim_{\omega\to 0} \lim_{k\to 0} (\omega/k^2)\, \chi_L''(k\omega) = \zeta + \frac{4}{3}\eta = L_L. \tag{23}$$

The poles of χ give the damped sound wave velocity and the thermal diffusivity. Likewise for $\chi_{nn}(k\omega)$ which equals

$$\frac{\dfrac{-nk^2}{m}\left[\omega + \dfrac{i}{mn}\dfrac{\varkappa}{c_v}k^2\right]}{\left[\omega + \dfrac{i}{mn}\dfrac{\varkappa}{c_p}k^2\right]\left[\omega^2 - \left(\dfrac{dp}{dn}\right)_s \dfrac{k^2}{m} + \dfrac{ik^2\omega}{mn}\left\{\dfrac{4}{3}\eta + \zeta + \varkappa\left(\dfrac{1}{c_v} - \dfrac{1}{c_p}\right)\right\}\right]}, \tag{24}$$

we have

$$\lim_{k\to 0} \chi_{nn}(k0) = n\left(\frac{dn}{dp}\right)_s \frac{c_p}{c_v} = n\left(\frac{dn}{dp}\right)_T, \tag{25}$$

$$\lim_{\omega\to 0} \lim_{k\to 0} (-\omega^2/k^2)\, \chi_{nn}(k\omega) = \frac{n}{m} = R_{nn}, \tag{26}$$

$$\lim_{\omega\to 0} \lim_{k\to 0} (\omega/k^2)\, \chi_{nn}''(k\omega) = 0 = L_{nn} \text{ (no diffusion).} \tag{27}$$

E. Measuring Correlation Functions by Scattering

Despite the use of the word measurement in the title of these lectures we have not spent much time referring to experimental methods. In this section we shall comment more extensively on how one measures some aspects of correlation functions. It would, of course, be delightful if all the external forces we talked about previously could be realized. Then we could apply them with arbitrary wavelength and frequency, measure the absorption and be done with it. In fact this is impossible for several reasons. First an external force like gravity which couples to the energy density couples too weakly. Secondly, even scalar interactions, which couple to the matter density cannot be effectively produced with an arbitrary wave number-frequency relation. There are, however, techniques that enable one to measure several important correlation functions – – those of density, electrical charge, current and magnetization. We shall consider them here paying special attention to those that measure the density correlation function and also the self or autocorrelation function, defined as.

$$\tilde{S}_{\text{self}}(\mathbf{r}\mathbf{r}';t-t') = \left\langle \sum_\alpha \delta(\mathbf{r}-\mathbf{r}_\alpha(t))\,\delta(\mathbf{r}'-\mathbf{r}_\alpha(t')) \right\rangle. \tag{1}$$

In a spatially invariant system we may write

$$
\begin{aligned}
S_{\text{self}}(k\omega) &= \int d(\mathbf{r}-\mathbf{r}')\,d(t-t')\,e^{-i\mathbf{k}\cdot(\mathbf{r}-\mathbf{r}')}\,e^{i\omega(t-t')}\,\tilde{S}_{\text{self}}(\mathbf{r}\mathbf{r}';t-t') \\
&= \int d(t-t')\left\langle \sum_\alpha e^{-i\mathbf{k}\cdot\mathbf{r}_\alpha(t)}\,e^{i\mathbf{k}\cdot\mathbf{r}_\alpha(t')}\,\delta(\mathbf{r}-\mathbf{r}_\alpha(t)) \right\rangle e^{i\omega(t-t')} \\
&= \int d(t-t')\,e^{i\omega(t-t')}\,\frac{1}{V}\left\langle \sum_\alpha e^{-i\mathbf{k}\cdot\mathbf{r}_\alpha(t)}e^{i\mathbf{k}\cdot\mathbf{r}_\alpha(t')} \right\rangle.
\end{aligned}
\tag{2}
$$

The associated autocorrelation functions, $\tilde{\chi}$, and $\tilde{\varphi}$ are defined in the usual way. These autocorrelation functions are closely related to the velocity autocorrelation functions we discussed in Section A. In particular, we have

$$\omega^2 S_{\text{self}}(k\omega) = \int dt\,e^{i\omega(t-t')}\left\langle \frac{1}{V}\sum_\alpha e^{-i\mathbf{k}\cdot\mathbf{r}_\alpha(t)}\mathbf{k}\cdot\dot{\mathbf{r}}_\alpha(t)\,e^{i\mathbf{k}\cdot\mathbf{r}_\alpha(t')}\mathbf{k}\cdot\dot{\mathbf{r}}_\alpha(t') \right\rangle$$

so that

$$
\begin{aligned}
\lim_{k\to 0}\frac{\omega^2 S_{\text{self}}(k\omega)}{k^2} &= \int dt\,e^{i\omega(t-t')}\left\langle \frac{1}{V}\sum_\alpha \hat{\mathbf{k}}\cdot\dot{\mathbf{r}}_\alpha(t)\,\hat{\mathbf{k}}\cdot\dot{\mathbf{r}}_\alpha(t') \right\rangle \\
&= \int dt\,e^{i\omega(t-t')}\left\langle \frac{1}{V}\sum_\alpha v_{\alpha 1}(t)\,v_{\alpha 1}(t') \right\rangle \\
&= n\int dt\,e^{i\omega(t-t')}\,\frac{1}{3}\langle \mathbf{v}(t)\cdot\mathbf{v}(t')\rangle = n S_{vv}(\omega)
\end{aligned}
\tag{3}
$$

is the velocity autocorrelation function and

$$
\begin{aligned}
\lim_{\omega\to 0}\lim_{k\to 0}\frac{\omega}{\beta k^2 n}\chi''_{\text{self}}(k\omega) &= \lim_{\omega\to 0}\frac{\chi''_{vv}(\omega)}{\beta\omega} = \lim_{\omega\to 0}\frac{1}{2\hbar\omega\beta}(1-e^{-\beta\omega\hbar})\,S_{vv}(\omega) \\
&= \frac{1}{2}S_{vv}(0) = [m\beta\gamma]^{-1}
\end{aligned}
\tag{4}
$$

is called the self-diffusion constant D_{self}.

We may measure functions $S_{nn}(k\omega)$ or $\chi''_{nn}(k\omega)$ for a medium by seeing how light waves or particle waves are scattered from it. The probing wave is used essentially as a billiard ball. The spectral function, $S_{nn}(k\omega)$, gives the number of density excitations of the system with a given energy and momentum. It is determined by measuring the number of times the billiard ball gains momentum $\hbar k$ and energy $\hbar\omega$ by bouncing off the system, and dividing by the known probability per unit density of excitations for absorbing such an excitation. The probability of losing energy and momentum by creating an excitation is related to the probability of absorbing one by a detailed balance argument.

Neutrons interact with the nuclei of atoms and molecules and with their magnetization. The nuclear and magnetic contributions can be separated experimentally. Photons interact directly with the charge, current, and magnetization. For neutral polarizable molecules, however, this interaction may be considered to be with the polarizability times the local density. Therefore both neutrons and photons can be regarded in simple liquids, as billiard balls which interact weakly with the density excitations, the kinematics of their interactions being depicted in Fig. 25a. For the photon, or x-ray, there

Fig. 25a. The kinematics of neutron scattering. For neutrons, the incident and outgoing energies, ε_i and ε_f are $p_i^2/2m_N$ and $p_f^2/2m_N$; for photons, $\varepsilon_i = \hbar\omega_i$ and $\varepsilon_f = \hbar\omega_f$ are $p_i c = \hbar k_i c$ and $p_f c = \hbar k_f c$.

are two experimental limitations, the wavelength of the photon, and the large energy per unit momentum imparted by a photon. As a result, the frequency or energy shift of the photon is so small that it is frequently regarded as having been scattered elastically. In wave terms, the frequency shift is essentially a Doppler shift which we know will be small. Nevertheless using lasers, which are very monochromatic sources, it is possible to resolve very small energy (or frequency) shifts. The important limitation is due to the fact that lasers use visible light and this imposes a severe wavelength limitation for fluids. For x-rays the wavelength limitation on resolution is not severe but the absence of monochromatic sources and of techniques for determining frequency shifts prevents one from measuring the frequency distribution of $S(k\omega)$. Only $(2\pi)^{-1}\int d\omega\, S_{nn}(k\omega) = \tilde{S}_{nn}(k, t = 0)$, the equilibrium correlation function can be measured.

Quantitatively, the maximum momentum transfer, twice the incident momentum, corresponds to a wavelength half the size of that of the incident photon. Experiments probe values of k between $2\omega/c$ and $(\omega/c)\,\theta$ where θ is the smallest resolvable angle. Characteristic energy losses are connected with the energy of phonons. For a given ω, $\Delta\omega \sim (c_{ph}/c)\,\omega\theta$. Thus the resolution required is $\Delta\omega/\omega \sim 10^{-5}$ in the backward direction and $10^{-5}\theta$, for small θ. This can be accomplished with a Fabry-Perrot interferometer. For excitations corresponding to temperature fluctuations the resolution must be even higher but with lasers homodyne (self-beating) techniques[44] can be used to detect a shift of a few cycles in 10^{15}.

With neutrons the resolution of short wavelengths poses no problem since momentum and energy imparted by a thermal neutron are quite large. The difficulty is one of collimating and monochromatizing neutrons and using them to measure small k. Small k corresponds to very small angles, that is, $k \sim (0.7 \times 10^9 E \text{ (in e.v.)})\,\theta$ cm^{-1}. In addition for small values of k, the energy change given by

$$\varepsilon_f - \varepsilon_i = \frac{\hbar^2}{2m_N}(\mathbf{k}_f + \mathbf{k}_i)\cdot(\mathbf{k}_f - \mathbf{k}_i) = \frac{\hbar^2}{2m_N}(2\mathbf{k}_i + \mathbf{k})\cdot\mathbf{k} \cong \frac{\hbar^2}{m_N}\mathbf{k}_i\cdot\mathbf{k}$$

is less than $\hbar c_{ph} k$ under all circumstances unless $c_{ph} < (\hbar k_i/m_N)$, that is unless $\varepsilon_i > m_N c_{ph}^2 \sim 2 \times 10^9 (c_{ph}/c)^2$ e.v. so that if $c_{ph}/c \sim 10^{-5}$, it is necessary to use neutrons with energies larger than ~ 0.1 e.v. to create long wavelength phonons.

Another practical difficulty should be mentioned. It is clear from our discussion that the convenient quantities to discuss are the frequency depend-

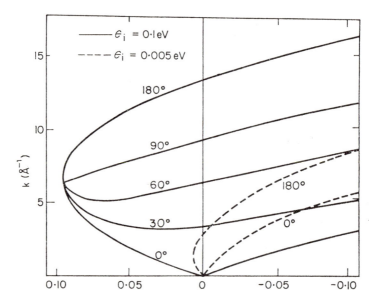

Fig. 25 b. Kinematics of neutron scattering. The wave vector transfer k as a function of the energy transfer $\hbar\omega$ for incident energies (ε_i) and several angles of scattering.

ent correlation functions for a fixed k. For photons, fixed k corresponds quite closely to fixed angle. With neutrons, however, fixed k experiments can be quite difficult. The reason is that to keep k fixed while varying the frequency over a sensible region, it is necessary to alter the incident energy (i.e. the fixed energy) and the scattering angle over extremely wide ranges. For example when $k \sim 1.0$ Å^{-1} and one wishes to measure all energy transfers less than 5×10^{-2} e.v., one must vary the incident neutron energy from 10^{-3} e.v. to 1 e.v. and the angle from $\sim 1°$ to $\sim 90°$. Figure 25b gives some feel for the kinematics in such experiments.[45]

We turn now to the demonstration[46] of the statement that experiments with photons and neutrons determine the function we called $S'_{nn}(k\omega)$, which in turn determines $\chi''_{nn}(k\omega)$ in equilibrium by the fluctuation dissipation theorem. In analyzing the neutron problem the interaction potential for nuclear scattering can be treated as weak and short range. More particularly we may replace $V(\mathbf{r}_N - \mathbf{r}^\alpha)$ by a point interaction $a\delta(\mathbf{r}_N - \mathbf{r}^\alpha)$. The quantity a is known as the pseudopotential and is actually spin dependent. If we neglect this spin dependence we may write the transition probability, the cross section times the incident flux, in the Born approximation (which is sufficiently accurate) as

$$\sigma \times \text{inc. flux} = \frac{2\pi}{\hbar} \sum_{\xi E} w_{\xi E} \int \frac{d\mathbf{k}_f}{(2\pi)^3} \sum_{\xi' E'} \delta(E - E' + \hbar\omega)$$

$$\times \left| \left\langle \mathbf{k}_f \xi' E + \hbar\omega \left| \sum_\alpha a\delta(\mathbf{r}_N - \mathbf{r}^\alpha) \right| \mathbf{k}_i \xi E \right\rangle \right|^2 \quad (5)$$

with

$$\hbar\omega = \frac{\hbar^2}{2m_N}(\mathbf{k}_f^2 - \mathbf{k}_i^2) \quad \text{and} \quad \mathbf{k} = \mathbf{k}_f - \mathbf{k}_i.$$

Introducing the density operator $n(\mathbf{r}) = \sum_\alpha \delta(\mathbf{r} - \mathbf{r}^\alpha)$ and the Fourier representation of the δ-function, we obtain

$$\sigma \times \text{inc. flux} = \int \frac{d\mathbf{k}_f}{(2\pi)^3} \int d\mathbf{r} \int d\mathbf{r}' \int \frac{d(t - t')}{\hbar^2} \sum_{\xi E} w_{\xi E} \sum_{\xi' E'} e^{-i\mathbf{k}\cdot(\mathbf{r}-\mathbf{r}') + i\omega(t-t')}$$

$$\times \exp[(E - E')(t - t')/i\hbar] \langle E'\xi' | a n(\mathbf{r}) | E\xi \rangle \langle E\xi | a^\dagger n(\mathbf{r}') | E'\xi' \rangle.$$

If we now recall that the Heisenberg operators satisfy

$$\left\langle E' \left| \exp\left(\frac{iE't'}{\hbar}\right) n(\mathbf{r}') \exp\left(\frac{-iEt'}{\hbar}\right) \right| E \right\rangle = \langle E' | n(\mathbf{r}'t') | E \rangle$$

and that the incident flux is $\hbar k_i/m_N$, we may write

$$\sigma = \frac{m_N}{\hbar^3 k_i} \int \frac{d\mathbf{k}_f}{(2\pi)^3} \int d\mathbf{r} \int d\mathbf{r}' \, e^{-i\mathbf{k}\cdot(\mathbf{r}-\mathbf{r}')} \int d(t - t') e^{i\omega(t-t')} \langle an(\mathbf{r}t) \, a^\dagger n(\mathbf{r}'t') \rangle.$$

8

The differential cross section is obtained by observing that

$$d\mathbf{k}_f = k_f^2 dk_f d\Omega_f = k_f^2 \frac{dk_f}{d\varepsilon_f} d\varepsilon_f d\Omega_f = \frac{m_N k_f}{\hbar^2} d\varepsilon_f d\Omega_f$$

$$\frac{d^2\sigma}{d\Omega_f d\varepsilon_f} = \frac{k_f m_N^2}{k_i \hbar^5 (2\pi)^3} \int d\mathbf{r} \int d\mathbf{r}'\, e^{-i\mathbf{k}\cdot(\mathbf{r}-\mathbf{r}')} \int d(t - t') e^{i\omega(t-t')} \langle an(\mathbf{r}t)\, a^\dagger n(\mathbf{r}'t')\rangle.$$

$$(6)$$

Up to this point we have not been completely candid about a which might be spin dependent; i.e., $a = a_1 + a_2 \boldsymbol{\sigma}_N \cdot \mathbf{S}_\alpha$.

Clearly if the term $a_2 \boldsymbol{\sigma}_N \cdot \mathbf{S}_\alpha$ is present, we really have

$$\langle an(\mathbf{r}t)\, a^\dagger n(\mathbf{r}'t')\rangle \to \langle a_1 n(\mathbf{r}t)\, a_1^* n(\mathbf{r}'t')\rangle + \langle a_2 \boldsymbol{\sigma}_N \cdot \mathbf{S}(\mathbf{r}t)\, a_1^* n(\mathbf{r}'t')\rangle$$

$$+ \langle a_1 n(\mathbf{r}t)\, a_2^* \boldsymbol{\sigma}_N \cdot \mathbf{S}(\mathbf{r}'t')\rangle + \langle a_2 \boldsymbol{\sigma}_N \cdot \mathbf{S}(\mathbf{r}t)\, a_2^* \boldsymbol{\sigma}_N \cdot \mathbf{S}(\mathbf{r}'t')\rangle$$

where $\mathbf{S}(\mathbf{r}t)$ is the nuclear spin density. Assuming the nuclear spins are uncorrelated, this expression reduces to

$$|a_1|^2 \langle n(\mathbf{r}t)\, n(\mathbf{r}'t')\rangle + |a_2|^2 S(S+1) \left\langle \sum_\alpha \delta(\mathbf{r} - \mathbf{r}_\alpha(t))\, \delta(\mathbf{r}' - \mathbf{r}_\alpha(t'))\right\rangle$$

$$= |\bar{a}|^2 \langle n(\mathbf{r}t)\, n(\mathbf{r}'t')\rangle + \left(\overline{|a|^2} - |\bar{a}|^2\right) \left\langle \sum_\alpha \delta(\mathbf{r} - \mathbf{r}_\alpha(t))\, \delta(\mathbf{r}' - \mathbf{r}_\alpha(t'))\right\rangle$$

We may therefore write $d^2\sigma/d\Omega_f d\varepsilon_f$ as the sum of two terms. The first is called the coherent scattering cross section

$$\frac{d^2\sigma_{\text{coh}}}{d\Omega_f d\varepsilon_f} = \frac{k_f}{k_i} \frac{m_N^2}{(2\pi)^3} \frac{1}{\hbar^5} |\bar{a}|^2 \int dt \int d\mathbf{r} \int d\mathbf{r}'\, e^{i\omega(t-t')-i\mathbf{k}\cdot(\mathbf{r}-\mathbf{r}')}$$

$$\times \left\{ \tilde{S}_{nn}(\mathbf{r}t;\mathbf{r}'t') + \langle n(\mathbf{r}t)\rangle \langle n(\mathbf{r}'t')\rangle \right\}.$$

$$(7)$$

The second term is called the incoherent scattering and is given by

$$\frac{d^2\sigma_{\text{inc}}}{d\Omega_f d\varepsilon_f} = \frac{k_f}{k_i} \frac{m_N^2}{(2\pi)^3} \frac{1}{\hbar^5} \left(\overline{|a|^2} - |\bar{a}|^2\right) \int dt \int d\mathbf{r} \int d\mathbf{r}'\, e^{i\omega(t-t')-i\mathbf{k}\cdot(\mathbf{r}-\mathbf{r}')}$$

$$\times \left\{ \tilde{S}_{\text{self}}(\mathbf{r}t;\mathbf{r}'t') \right\}.$$

$$(8)$$

In a solid, the second term in (7) contributes to elastic Bragg scattering; in a fluid, it only contributes at zero angle. The first term gives in a fluid for the cross section per volume of fluid

$$\frac{d^2\sigma_{\text{coh}}}{V d\Omega_f d\varepsilon_f} = \frac{k_f}{k_i} \frac{m_N^2}{(2\pi)^3} \frac{|\bar{a}|^2}{\hbar^5} S_{nn}(k\omega).$$

$$(9)$$

This would be the entire answer if the nuclear scattering amplitude a were spin-independent. In addition we have the incoherent contribution (8) depending on $\overline{|a|^2} - |\bar{a}|^2$

$$\frac{d^2\sigma_{\text{inc}}}{Vd\Omega_f d\varepsilon_f} = \frac{k_f}{k_i} \frac{m_N^2}{(2\pi)^3} \frac{(\overline{|a|^2} - |\bar{a}|^2)}{\hbar^5} S_{\text{self}}(k\omega). \tag{10}$$

Some typical results[47] for $S(k\omega)$ and $S_{\text{self}}(k\omega)$ can be seen in Fig. 26.

Clearly the same arguments can be applied to electron scattering off a charge distribution. Indeed we need only substitute

$$a\delta(\mathbf{r}_N - \mathbf{r}^\alpha) \to \frac{ee^\alpha}{4\pi |\mathbf{r} - \mathbf{r}^\alpha|}, \quad a \to \frac{ee^\alpha}{k^2}, \quad \text{and} \quad m_N \to m_e \tag{11}$$

(note that we have used *rationalized* Gaussian units) so that for a uniform system the cross section per unit volume is

$$\frac{d^2\sigma}{Vd\Omega_f d\varepsilon_f} = \frac{k_f}{k_i} \frac{m_e^2}{(2\pi)^3} \frac{e^2}{\hbar^5 k^4} S_{\varrho\varrho}(k\omega) \tag{12}$$

(ϱ is the charge density).

In a neutral system, in which the charge distribution is localized on atoms which are polarized by photons we have a similar formula, expressing the fact that polarization fluctuations give rise to inelastic photon scattering. A more critical examination of the derivation of this formula than those in the literature ought certainly be undertaken. To the extent that one treats the polarizability as a local quantity depending on the density, however, the derivation is straightforward. A classical calculation of the effect of these polarization fluctuations gives for electromagnetic scattering[48]

$$\frac{d^2\sigma}{Vd\Omega_f d\omega_f} = \left(\frac{1}{2} \frac{\partial\varepsilon}{\partial n}\right)^2 \left(\frac{\omega_i}{c}\right)^4 \frac{(\mathbf{e}_i \times \hat{\mathbf{k}}_f)^2}{(2\pi)^3} S_{nn}(k\omega)\sqrt{\varepsilon} \tag{13}$$

where $\mathbf{e}_i$ is the incident polarization vector. If we have a system for which $\hbar\omega << \varepsilon_i$ for all significant contributions to the scattering, it is quasi-elastic and the cross section for neutrons is

$$\frac{d\sigma_{\text{coh}}}{Vd\Omega} = \int \frac{d^2\sigma_{\text{coh}}}{d\Omega d\varepsilon_f} d\varepsilon_f = m_N^2 \frac{|\bar{a}|^2}{(2\pi)^2 \hbar^4} \int \frac{d\omega}{2\pi} S_{nn}(k\omega) = \frac{m_N^2 |\bar{a}|^2}{(2\pi)^2} \frac{\tilde{S}_{nn}(k)}{\hbar^4}$$

while for photons it is

$$\frac{d\sigma_{\text{coh}}}{Vd\Omega} = \frac{1}{4}\left(\frac{\omega_i}{c}\right)^4 \frac{(\mathbf{e}_i \times \hat{\mathbf{k}}_f)^2}{(2\pi)^2} \tilde{S}_{nn}(k)\left(\frac{\partial\varepsilon}{\partial n}\right)^2 \sqrt{\varepsilon} \tag{14}$$

where $\tilde{S}_{nn}(k)$ is the instantaneous correlation function. This approximation will always apply even for X-ray scattering unless unforeseen resolution is attained.

While we shall not give a satisfactory derivation of the above Brillouin scattering formula let us at least make it plausible by noting that for small k,

8*

for a rare gas $S_{nn}(k\omega) = 2n\delta(\omega)\,\pi$. Then we have $\varepsilon - 1 = n\alpha$ and

$$\frac{d\sigma}{V d\Omega} = \frac{n\alpha^2}{(4\pi)^2} \left(\frac{\omega_i}{c}\right)^4 (\mathbf{e}_i \times \hat{\mathbf{k}}_f)^2 \tag{15}$$

where α is the polarizability. This is just what we should find, that is,

$$\frac{d\sigma}{d\Omega} = N \left[\frac{\overline{\ddot{\mathbf{p}}(t)^2}}{(4\pi)^2 c^3} (\mathbf{e}_i \times \hat{\mathbf{k}}_f)^2 \right] [c\overline{\mathbf{E}_{ext}^2(t)}]^{-1}$$

$$= N \begin{bmatrix} \text{power radiated per dipole} \\ \text{per steradian} \end{bmatrix} \text{[incident power per unit area]}^{-1}.$$

As a final example of a scattering experiment for determining a correlation function we return to neutrons. Although they are uncharged, neutrons

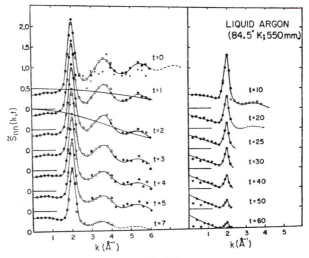

Fig. 26a

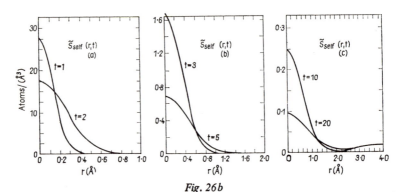

Fig. 26b

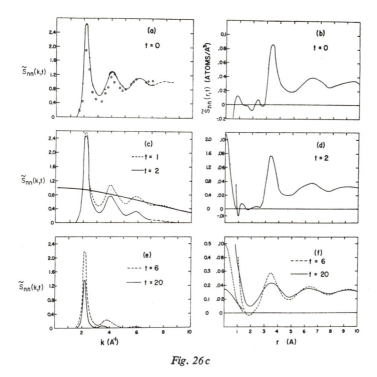

Fig. 26c

Fig. 26. Some typical structure functions determined by neutron scattering in liquid argon and liquid lead. (a) $\tilde{S}_{nn}(kt)$ in argon for different times in units of 10^{-13} sec. The circles are obtained from the experimental $S_{nn}(k\omega)$ by Fourier transformation; (b) the self correlation function in argon, $\tilde{S}_{\text{self}}(rt)$, in the same units; (c) $\tilde{S}_{nn}(kt)$ and $\tilde{S}_{nn}(rt)$ for liquid lead. In the last two graphs, $\tilde{S}_{\text{self}}(rt)$ is also plotted; its scale is the left one.

possess a magnetic moment. This magnetic moment interacts with the magnetization of the medium by the magnetic dipole interaction. As a result, there will be scattering once again, and again the scattering will be given by a correlation function. If we describe the magnetization by $S_{\text{MM}} = |F(k)|^2 S_{ss}$ the matrix element for magnetic neutron scattering is given by the Biot-Savart law

$$-M \cdot H \rightarrow \left(1.91 \frac{e\hbar}{m_N c}\right)\left(\frac{e\hbar}{m_e c}\right)[(S^\alpha \cdot \hat{\mathbf{k}})(S^N \cdot \hat{\mathbf{k}}) - (S^\alpha \cdot S^N)]\, F(k). \quad (16)$$

The quantity $F(k)$ is a magnetic form factor. Assuming we can average over directions, we may write the cross section per unit volume as

$$\frac{d^2\sigma_{\text{Mag}}}{V d\Omega_f d\varepsilon_f} = \frac{k_f}{k_i} \frac{1}{(2\pi)^3}\left(1.91 \frac{e^2}{m_e c^2}\right)^2 \frac{1}{6\hbar} S_{\text{s}\cdot\text{s}}(k\omega)\, |F(k)|^2 \quad (17)$$

TABLE I

	Radiation	Target	Data type	Momentum range ($\hbar k$)	Energy range ($\hbar\omega$)	Interpretation	Isotopic variation	Metal samples
S_{self}, S_{nn} 1	Neutron scattering	Nucleus	Nuclear position correlation	Wide	Wide	Straight-forward	Yes	Yes
S_{self} 2	Neutron absorption	Nucleus	Nuclear position correlation	Single value	Wide	Straight-forward	Yes	Yes
S_{nn} 3	Electron scattering	Charge cloud	Atomic position correlation	Wide	Integrated	Form factor	No	Yes
S_{nn} 4	X-ray scattering	Electron cloud	Atomic position correlation	Wide	Integrated	Form factor	No	Yes
S_{nn} 5	γ-ray scattering	Electron cloud	Atomic position correlation	Wide	Very narrow	Form factor	No	Yes
S_{self} 6	γ-ray absorption/emission	Nucleus	Nuclear position correlation	Single value	Very narrow	Simple	Rare	Yes
S_{nn} 7	Raman scattering	Electron cloud	Atomic position correlation	Small	Wide	Selection rules	No	No
S_{nn} 8	Brillouin scattering	Electron cloud	Atomic position correlation	Small	Small	Selection rules	No	No
S_{nn} 9	Ultrasonic transmission	Local density	Density correlation	Very small	Very small	Straight-forward	No	Yes

A number of these techniques determine other correlations as well (e.g. neutrons measure magnetization correlations; electrons measure charge correlations, even when the charge is not coupled to the atomic density as it is in insulators). Moreover other techniques, notably infrared absorption and emission, dielectric relaxation, and nuclear magnetic resonance, measure magnetic and electric polarization correlations, which under some circumstances are related to the density or self-correlation function.

in terms of the trace of the spin correlation function $S_{s\text{-}s}$. Note again, that there is an extra $(4\pi)^2$ relative to the conventional formula because we have used rationalized units, and also a difference of a factor of the density because our correlation functions are defined without the factor of $1/N$, and our cross sections are per volume.

When the inelasticity is negligible we obtain

$$\frac{d\sigma_{\text{Mag}}}{V d\Omega} = \tilde{S}_{s\text{-}s}(k)\left(\frac{1.91\ e^2}{4\pi m_e c^2}\right)^2 \frac{2}{3}\ |F(k)|^2 \tag{18}$$

and for independent point particles, with $S_{s\text{-}s}(k) \sim S(S+1)\,n$, and $F(k) = 1$, the cross section is

$$\frac{d\sigma_{\text{Mag}}}{d\Omega} = \frac{2S(S+1)\,N}{3}\left(\frac{1.91\ e^2}{4\pi m_e c^2}\right)^2. \tag{19}$$

Just as $S_{nn}(k\omega) = 2\hbar\chi''_{nn}(k\omega)\,[1 - e^{-\beta\omega\hbar}]^{-1}$ is large in the neighborhood of a liquid gas critical point, giving rise to the phenomenon of critical opalescence $S_{s\text{-}s}(k\omega) = 2\hbar\chi''_{s\text{-}s}(k\omega)\,[1 - e^{-\beta\omega\hbar}]^{-1}$ is large for $k = 0$ at a ferromagnetic transition and for $k \sim$ reciprocal lattice vector at an antiferromagnetic transition.[49] Phenomenologically, it would also be expected to narrow, i.e. the fluctuations should slow down. Although it has proven difficult to verify this expectation the most recent experiments[50] leave little doubt that narrowing occurs in antiferromagnets at the appropriate value of k.

The experimental techniques we have discussed in this section are the most searching probes of the correlation functions. But, as we have seen previously, transport and thermodynamic measurements study these same functions at a single point. Ultrasonic measurements trace out the values of the correlation functions on a line or set of lines in the k, ω plane. That is to say they describe $\chi''(k\omega)$ near resonances $\bar{\omega}(k)$. The reader, in other words, will find if difficult to think of experiments whose results are not given in terms of some simple correlation function of the type we have been discussing. Conversely a table of what parts of the density correlation function are provided by different scattering techniques may be useful. Such a table has been compiled by Egelstaff and is included here.[51]

F. Correlations in Charged Systems

Of the correlation functions occurring in solid state physics, perhaps the most frequently encountered are those of the charge density, and current. We have already remarked on the fact that the former can be measured by inelastic electron scattering. The latter is more difficult to measure under varied circumstances but is clearly relevant to optical absorption experiments, for example. From a theoretical point of view, these correlation functions also serve an instructive purpose; they show, by counterexample, that a number of the properties we discussed previously are not completely trivial. The standard expressions for transport coefficients, for example, do not hold in charged systems. The reason they do not stems from the long range nature of electromagnetic forces.[52]

More specifically the transport coefficients relate the currents to the total electric field and the relation between total field and applied field which was proven in the long wavelength limit for uncharged systems does not hold because the quantity corresponding to a thermodynamic derivative in the uncharged case, $\partial \mu / \partial n$ here corresponds to $\partial \varphi / \partial \varrho \approx k^2 \to 0$ as $k \to 0$.

1. Macroscopic Equations

Before departing on our correlation function discussion, let us briefly comment on the macroscopic Maxwell equations — equations which in many ways are analogous to the hydrodynamic equations of a simple fluid.

The quantities which occur in the macroscopic equations for a simple fluid are averages of the microscopically well defined densities and currents. The macroscopic equations for the electromagnetic fields in media are less defined. The microscopic current is represented by three macroscopic quantities when it varies slowly in space and time; i.e.

$$\langle \mathbf{J} \rangle \to \langle \mathbf{J}^{\text{free}} \rangle + \left\langle \frac{\partial \mathbf{P}}{\partial t} \right\rangle + \langle c \nabla \times \mathbf{M} \rangle$$

but the nature of this "tripartite" separation is poorly defined; neither the division into bound and free charge, nor the division of the bound charge into a polarization and a magnetization has an unambiguous meaning. Since electrons are indistinguishable and exchange collisions take place we can clearly not separate charges and currents into free parts and bound parts. Likewise, as is stressed in the more honest, more recent college texts (e.g. the Berkeley series[53]) the division of bound charge into polarization and magnetization is ambiguous even in some of the simplest situations.

Despite this ambiguity, there is a less ambitious practical separation which conforms with a great deal of the usage of the macroscopic equations. At least in linear isotopic media, (and also somewhat more generally), we may effectively divide current matrix elements, into different parts, i.e., define three different coefficients, a conductivity, an electric susceptibility, and a magnetic susceptibility, each of which gives rise to a current with a different space time dependence in the long wavelength limit. The three parts of the response, in a restricted sense, may then be said to give meaning in the linear regime to the free current, the polarization, and the magnetization.

Specifically we may write

$$J_i \rightarrow \sigma'_{ij}E_j - i\omega(\varepsilon' - 1)_{ij}E_j + \int \varepsilon_{ijk}\nabla_j(1 - \mu^{-1})_{kl}(\mathbf{rr}')\frac{c^2}{i\omega}\varepsilon_{lmn}\nabla_m E_n(\mathbf{r}')\,d\mathbf{r}' \tag{1}$$

where σ'_{ii} is the real conductivity, ε'_{ij} the real dielectric constant, μ_{ij} the real permeability, and ε_{ijk} the Levi-Civita symbol. Note that the terms involving the electric and magnetic permeabilities do not appear in the hydrodynamic description when we let $k \rightarrow 0$ and then $\omega \rightarrow 0$. In that limit, only the constant term remains in the conductivity. Here, as elsewhere, the possibility of a leading reactive term, proportional to $1/i\omega$, should not be ignored. We shall use this term, $\dot{J}_i \equiv (\varrho_s/\varrho)_{ij}\,\omega_p^2 E_i$ where ω_p^2 is the plasma frequency, to define the superconductive parameter $(\varrho_s/\varrho)_{ij}$.

While the theorems that apply for neutral systems are modified, we can still apply an external field, and moreover, the induced current is still the current correlation function times the external field. However, we must now correct for the difference between the externally applied field and the total field even in the long wavelength low frequency limit. By making this correction we shall obtain formulas for the transport coefficients. Before turning to their derivation let us repeat the other distinction between the macroscopic coefficients in these equations and hydrodynamic parameters. As is apparent from Eq. (1), the static electric and magnetic permeabilities are of higher order in the frequency and wave number than the corresponding hydrodynamic expressions, i.e., they are related to the last two terms in the derivative expansion

$$\dot{J}_i = (\varrho_s/\varrho)_{ij}\,\omega_p^2 E_j + \sigma'_{ij}\dot{E}_j + (\varepsilon' - 1)_{ij}\,\ddot{E}_j + \bar{\gamma}_{ijkl}\,\nabla_j\nabla_k E_l. \tag{2}$$

Conservation laws therefore play a lesser role in their discussion than in the truly long wavelength and low frequency terms which describe the superfluid density and conductivity.

Note incidentally that we may also look upon the first or superconducting term as a diamagnetic contribution to $\bar{\gamma}$ and hence the magnetization, which is of infinite range (i.e. $\bar{\gamma} \sim 1/\nabla^2$). This ambiguity will find a natural counterpart in our correlation function discussion.

Note finally that once we allow for wave number and frequency dependent response functions the entire current response to the electromagnetic field really involves, as does the unseparated (and inseparable) microscopic current, a single tensor transport coefficient which we may call the conductivity, the dielectric constant, or whatever we wish. It is this one tensor relation which has a well defined microscopic meaning and consequently a correlation function counterpart before we make a wavelength and frequency expansion. Like the tensor response function for the momentum in a simple fluid, it can be divided in an isotopic medium into two scalar response functions, a transverse and longitudinal response. It is these two coefficients for which we shall find correlation function expressions.

2. Current Response

One formal aspect of the correlation function description concerns the fact that the current, in the presence of an electromagnetic field, is given by

$$\mathbf{J} = \sum_i \frac{e^i}{2m^i} \sum_\alpha \left\{ \mathbf{p}_\alpha^i - \frac{e^i}{c} \mathbf{A}(\mathbf{r}t), \delta(\mathbf{r} - \mathbf{r}_\alpha^i(t)) \right\}$$
$$+ c\nabla \times \sum_i \mu^i \sum_\alpha (\mathbf{s}_\alpha^i/s^i) \delta(\mathbf{r} - \mathbf{r}_\alpha^i(t)) \qquad (4)$$

where i refers to the species and α to the particle of that species; e^i, m^i and μ^i are the charge, mass and magnetic moment of particles of the ith species. The current operator depends explicitly on the potential $\mathbf{A}$ and our formula for induced changes in an operator involves only the implicit change. Thus if we change the hamiltonian by applying an external vector potential $\mathbf{A}^{ext}$, and an external scalar potential φ^{ext}, we have when we include the explicit change in $\mathbf{J}^{ind}$, the induced current,

$$\delta\langle J_k^{ind}(\mathbf{r}t)\rangle = \int \tilde{\chi}_{J_k J_l}(\mathbf{r}\mathbf{r}'; t - t') \frac{1}{c} \delta A_l^{ext}(\mathbf{r}'t')$$
$$- \int \tilde{\chi}_{J_k \varrho}(\mathbf{r}\mathbf{r}'; t - t') \delta\varphi^{ext}(\mathbf{r}'t') - \frac{\omega_p^2}{c} \delta A_k^{ext}(\mathbf{r}t) \qquad (5)$$

where $\omega_p^2 = \sum_i \langle n^i \rangle e^{i2}/m^i$, is the plasma frequency (again, in rationalized units). Let us first observe that this expression is gauge invariant.[54] Thus if we perform the transformation,

$$\bar{\varphi}^{ext}(\mathbf{r}t) = \varphi^{ext}(\mathbf{r}t) - \frac{1}{c} \frac{\partial}{\partial t} \Lambda(\mathbf{r}t)$$

$$\bar{\mathbf{A}}^{ext}(\mathbf{r}t) = \mathbf{A}^{ext}(\mathbf{r}t) + \nabla\Lambda(\mathbf{r}t) \qquad (6)$$

the formula for δJ_k^{ind} is altered by

$$- \omega_p^2 \frac{1}{c} \nabla_k \Lambda + 2i \int_{-\infty}^{t} dt' d\mathbf{r}' \, \tilde{\chi}_{J_k J_l}''(\mathbf{r}\mathbf{r}'; t - t') \frac{1}{c} \nabla_l' \Lambda(\mathbf{r}'t')$$

$$+ 2i \int_{-\infty}^{t} dt' d\mathbf{r}' \, \tilde{\chi}_{J_k \varrho}''(\mathbf{r}\mathbf{r}'; t - t') \frac{1}{c} \frac{\partial}{\partial t'} \Lambda(\mathbf{r}'t'). \qquad (7)$$

Integrating by parts with respect to time and space we obtain

$$- \omega_p^2 \frac{1}{c} \nabla_k \Lambda + \int 2i \tilde{\chi}''_{J_k \varrho}(\mathbf{rr'} ; 0) \frac{1}{c} \Lambda(\mathbf{r'}t) \, d\mathbf{r'}$$

$$- \int_{-\infty}^{t} d\mathbf{r'} dt' 2i \left[\nabla'_l \tilde{\chi}''_{J_k J_l}(\mathbf{rr'} ; t - t') + \frac{\partial}{\partial t'} \tilde{\chi}''_{J_k \varrho}(\mathbf{rr'} ; t - t') \right] \frac{1}{c} \Lambda(\mathbf{r'}t'). \qquad (8)$$

The second line vanishes by current conservation; the first line by the now familiar equal time commutation relation

$$-i \nabla_k \omega_p^2 \, \delta(\mathbf{r} - \mathbf{r'}) = \sum_l \langle [j_k^l(\mathbf{r}t), \varrho^l(\mathbf{r'}t)] \rangle. \qquad (9)$$

We may therefore with no loss in generality choose the gauge in which $\varphi^{\text{ext}} = 0$, and write the Fourier transform of the induced current in the form

$$J_k^{\text{ind}}(\mathbf{r}\omega) = \int d\mathbf{r'} \, [\chi_{J_k J_l}(\mathbf{rr'} ; \omega) - \omega_p^2 \, \delta(\mathbf{r} - \mathbf{r'}) \delta_{kl}] \, E_l^{\text{ext}}(\mathbf{r'} \, \omega)/i\omega. \qquad (10)$$

Although this expression is gauge invariant, it does not characterize the electromagnetic transport properties of the system in the most conventional manner. Ordinarily, the conductivity of a system is not defined as the coefficient which relates the induced current to the externally applied field, but as the coefficient relating the induced current to the total electric field in the system. To convert (10) into a relation between the induced current and the total field in the system,[55] we write in a schematic matrix form

$$\int d\mathbf{r'} \left[\nabla^2 \mathbf{1} - \nabla\nabla + \frac{\omega^2}{c^2} \mathbf{1} \right] (\mathbf{rr'}) \, \delta\mathbf{E}(\mathbf{r'}) = \frac{1}{c^2} \delta\mathbf{J}^{\text{tot}} = \frac{1}{c^2} \delta\mathbf{J}^{\text{ind}} + \frac{1}{c^2} \delta\mathbf{J}^{\text{ext}}$$

$$= \int d\mathbf{r'} \left[\nabla^2 \mathbf{1} - \nabla\nabla + \frac{\omega^2}{c^2} \mathbf{1} \right] (\mathbf{rr'}) \, \delta\mathbf{E}^{\text{ext}}(\mathbf{r'})$$

$$- \frac{1}{c^2} \int d\mathbf{r'} \, [\chi_{JJ} - \omega_p^2 \mathbf{1}] \, (\mathbf{rr'} ; \omega) \, \delta\mathbf{E}^{\text{ext}}(\mathbf{r'}). \qquad (11)$$

In other words we use (10) and the relation between the external current and the external field to find the total field $\delta\mathbf{E}(\mathbf{r})$ in terms of the external field, $\delta\mathbf{E}^{\text{ext}}$. Once we have this relation

$$\delta\mathbf{E}^{\text{ext}} = \left[\nabla^2 \mathbf{1} - \nabla\nabla + \frac{\omega^2}{c^2} \mathbf{1} - \frac{1}{c^2} \chi_{JJ} + \frac{\omega_p^2}{c^2} \mathbf{1} \right]^{-1} \left[\nabla^2 \mathbf{1} - \nabla\nabla + \frac{\omega^2}{c^2} \mathbf{1} \right] \delta\mathbf{E}$$

$$\qquad (12)$$

we may substitute in (10) obtaining

$$\delta\mathbf{J}^{\text{ind}}(\mathbf{r}) = -i\omega \int \sigma \, (\mathbf{rr'} ; \omega) \, d\mathbf{r'} \, \delta\mathbf{E} \, (\mathbf{r'}) \qquad (13)$$

where

$$i\omega\boldsymbol{\sigma} = (\boldsymbol{\chi_{JJ}} - \omega_p^2) + (\boldsymbol{\chi_{JJ}} - \omega_p^2)$$
$$\times \left[\nabla^2 \mathbf{1} - \nabla\nabla + \frac{\omega^2}{c^2}\mathbf{1} - \frac{1}{c^2}\boldsymbol{\chi_{JJ}} + \frac{\omega_p^2}{c^2}\mathbf{1} \right]^{-1} \frac{(\boldsymbol{\chi_{JJ}} - \omega_p^2)}{c^2}. \quad (14)$$

In an isotropic medium which is invariant under time reversal, we may write

$$\chi_{J_i J_j} = \chi_L \frac{k_i k_j}{k^2} + \chi_T \left(\frac{k^2 \delta_{ij} - k_i k_j}{k^2} \right) \quad (15)$$

$$\sigma_{ij} = \sigma_L \frac{k_i k_j}{k^2} + \sigma_T \left(\frac{k^2 \delta_{ij} - k_i k_j}{k^2} \right), \quad (16)$$

and

$$\omega^2 \chi_{\varrho\varrho} = k^2(\chi_L - \omega_p^2). \quad (17)$$

In that case the formulas reduce to

$$\varepsilon_L - 1 \equiv \frac{i\sigma_L}{\omega} = \frac{\chi_L - \omega_p^2}{\omega^2}\left[1 - \frac{\chi_L - \omega_p^2}{\omega^2} \right]^{-1} = \frac{\chi_{\varrho\varrho}}{k^2}\left[1 - \frac{\chi_{\varrho\varrho}}{k^2} \right]^{-1}$$

$$\varepsilon_L = 1 + \frac{i\sigma_L}{\omega} = \left[1 - \frac{\chi_{\varrho\varrho}}{k^2} \right]^{-1} \quad (18)$$

$$\varepsilon_T - 1 \equiv \frac{i\sigma_T}{\omega} = \frac{\chi_T - \omega_p^2}{\omega^2}\left[1 - \frac{\chi_T - \omega_p^2}{\omega^2 - c^2 k^2} \right]^{-1} \quad (19)$$

the transverse projection operator, $\nabla^2 - \nabla\nabla$ disappearing from (18).

3. Longitudinal Response

Because the longitudinal response is essentially the response of the charge density a conserved quantity, the description of its long wavelength limit (in which only the "free charge" participates) conforms with other hydrodynamic discussions. The "thermodynamic sum rule" still holds but

$$\left(\frac{d\varrho}{d\varphi} \right)_{p,\,T} = \lim_{k\to 0} \int \frac{d\omega}{\pi} \frac{\chi_{\varrho\varrho}''(k\omega)}{\omega} = k^2 \to 0 \quad (20)$$

which indicates where the analysis of charged systems analogous to the earlier discussion, fails. A more exact statement of this sum rule (which reflects the screening of the electric field) is

$$1 = \lim_{k\to 0} \int \frac{d\omega}{\pi} \frac{\chi_{\varrho\varrho}''(k\omega)}{\omega k^2} = \lim_{k\to 0} \int \frac{d\omega}{\pi} \frac{\chi_L''(k\omega)}{\omega^3} = -\lim_{k\to 0} \int \frac{d\omega}{\pi} \frac{1}{\omega} \operatorname{Im} \frac{1}{\varepsilon_L(k\omega)}$$

$$= \lim_{k\to 0} \int \frac{d\omega}{\pi} \frac{1}{\omega^2} \frac{\sigma_L'(k\omega)}{|\varepsilon_L(k\omega)|^2}. \quad (21)$$

The high frequency sum rule for χ'', which corresponds to (9) is

$$\omega_p^2 = \int \frac{d\omega}{\pi} \frac{\chi_L''(k\omega)}{\omega} = \int \frac{d\omega}{\pi} \frac{\chi_{\varrho\varrho}''(k\omega)\,\omega}{k^2} = -\int \frac{d\omega}{\pi} \omega \mathrm{Im} \frac{1}{\varepsilon_L(k\omega)}$$

$$= \int \frac{d\omega}{\pi} \frac{\sigma_L'(k\omega)}{|\varepsilon_L(k\omega)|^2}. \tag{22}$$

Let us note also that *if*

$$\varepsilon_L^{-1} = \left(\frac{i\sigma_L}{z} + 1\right)^{-1} = 1 - \frac{\chi_{\varrho\varrho}(kz)}{k^2} \tag{23}$$

has no zeros for complex values of z, ε_L is also analytic and we may write

$$\varepsilon_L(kz) - 1 = \int \frac{d\omega}{\pi} \frac{\varepsilon_L''(k\omega)}{\omega - z} = \int \frac{d\omega}{\pi} \frac{\omega\varepsilon_L''(k\omega)}{\omega^2 - z^2} \tag{24}$$

$$= \int \frac{d\omega}{\pi} \frac{\sigma_L'(k\omega)}{\omega^2 - z^2}. \tag{25}$$

From this assumption of analyticity there follow two additional sum rules,

(a) $\quad\quad \varepsilon_L(k0) - 1 = \int \frac{d\omega}{\pi} \frac{\varepsilon_L''(k\omega)}{\omega}$ is real and positive; $\quad\quad$ (26)

and since $\varepsilon_L(kz) - \underset{z \to \infty}{1} = \chi_{\varrho\varrho}(kz)/k^2 = \underset{z \to \infty}{-\omega_p^2/z^2}$,

(b) $\quad\quad \int \frac{d\omega}{\pi} \sigma_L'(k\omega) = \omega_p^2 \quad \left(= \int \frac{d\omega}{\pi} \omega\varepsilon_L''(k\omega)\right).$ $\quad\quad$ (27)

As we shall see at a later stage, the functions σ_L' and $-\omega\,\mathrm{Im}\,\varepsilon_L^{-1} = \sigma_L'|\varepsilon_L|^{-2}$ which have the same frequency integral, are extremely different for free charge at long wavelengths. Specifically, the strength of the latter lies at the plasma frequency, $\pm\omega_p$

$$-\omega\mathrm{Im}\,\varepsilon_L^{-1} = \omega \frac{\varepsilon_L''}{(\varepsilon_L')^2 + (\varepsilon_L'')^2} \cong \pi\omega_p^3\delta(\omega^2 - \omega_p^2) \tag{28}$$

$$\frac{1}{\varepsilon_L} \cong \frac{\omega^2}{\omega^2 - \omega_p^2} \tag{29}$$

while the strength of the former which is essentially the conductivity, lies at low frequencies

$$\sigma_L'(\omega) = \omega\varepsilon_L''(\omega) \cong \pi\omega_p^2\delta(\omega); \ 1 + \frac{i\sigma_L}{\omega} = \varepsilon_L \cong 1 - \frac{\omega_p^2}{\omega^2}. \tag{30}$$

When collisions are taken into account the latter is significantly modified

$$\sigma'_L(\omega) = \omega \varepsilon''_L(\omega) \cong \omega_p^2 \frac{\tau}{1 + (\omega\tau)^2} \; ; \; \varepsilon_L \cong 1 - \frac{\omega_p^2}{\omega(\omega + i/\tau)} \qquad (31)$$

whereas the former is not. When $k \to 0$, $\varepsilon_L = 1 + i\sigma_L/\omega$ diverges as $(\varepsilon_0 k_s^2/k^2)$, where ε_0 and k_s^2 are defined by

$$\lim_{k\to 0}(\varepsilon_L(k0) - 1) = \lim_{k\to 0}\int \frac{d\omega}{\pi}\frac{\varepsilon''_L(k\omega)}{\omega} \equiv \frac{\varepsilon_0 k_s^2}{k^2} + (\varepsilon_0 - 1) + 0(k^2) \qquad (32)$$

The quantities have been defined in this way since, when a small k expansion is sufficient, we obtain for the potential $\varphi(\mathbf{r})$ of a static charge

$$\varphi(\mathbf{r}) = \int \frac{d\mathbf{k}}{(2\pi)^3} e^{i\mathbf{k}\cdot\mathbf{r}} \frac{1}{k^2\varepsilon(k0)} \cong \int \frac{d\mathbf{k}}{(2\pi)^3} e^{i\mathbf{k}\cdot\mathbf{r}} \frac{1}{\varepsilon_0(k^2 + k_s^2)} = \frac{e^{-k_s r}}{4\pi\varepsilon_0 r}. \qquad (33)$$

Thus k_s^{-1} gives the screening length of a charge when this expansion is valid. It should be stressed that the Fourier transform of the small k expansion of ε does not necessarily give the asymptotic behavior of the potential φ for large r. It can be particularly deceptive when $\varepsilon(k)$ is singular[56] for some other real value of k.

Combining the small k and ω dependence we infer that at long wavelengths and low frequencies the characteristic behavior of ε_L is

$$\varepsilon_L - 1 \cong \frac{\omega_p^2 \tau}{- i\omega + k^2 u^2 \tau - \omega^2 \tau} \qquad (34)$$

or

$$\frac{\sigma'_L(k\omega)}{\omega} = \varepsilon''_L(k\omega) \cong \omega_p^2(\omega/\tau)\,[(\omega^2 - k^2 u^2)^2 + (\omega/\tau)^2]^{-1} \qquad (35)$$

with

$$u^2 = \omega_p^2/k_s^2\varepsilon_0. \qquad (36)$$

Note now that as $k \to 0$ this discussion can be made quite parallel to that of the previous sections. There we showed that all the properties we have assumed for σ_L are in fact true for $\bar{\sigma}_L$ defined by

$$\left[1 - \frac{\chi_{\varrho\varrho}(kz)}{\chi_{\varrho\varrho}(k0)}\right] \equiv \left[1 + \frac{i\bar{\sigma}_L(kz)}{z}\right]^{-1} = z^2\left[z^2 + iz\int \frac{d\omega}{\pi}\frac{\bar{\sigma}'_L(k\omega)}{\omega^2 - z^2}\right]^{-1} \qquad (37)$$

(where $\bar{\sigma}'_L$ may have an irregular term like Γ in Sec. B) but as $k \to 0$, $\chi_{\varrho\varrho}(k0) \to k^2$ so that $\lim_{k\to 0}\bar{\sigma}_L(kz) = \lim_{k\to 0}\sigma_L(kz)$, and in this limit the equation

$$\lim_{k\to 0}\int \frac{d\omega}{\pi}\bar{\sigma}'_L(k\omega) = \lim_{k\to 0}\int \frac{d\omega}{\pi}\sigma'_L(k\omega) = \omega_p^2 \qquad (38)$$

is just the usual rule (52) discussed in Section B. Indeed the corresponding interpolation formula for $\mathrm{Im}\, z > 0$

$$\lim_{k \to 0} \bar{\sigma}'_L(kz) = \lim_{k \to 0} \sigma_L(kz) = \frac{\omega_p^2 \tau}{1 - iz\tau} \tag{39}$$

is hardly a new one. For larger k, $\bar{\sigma}_L(kz)$ must satisfy all the usual equations, (those proven for $\Gamma(z)$ in section B), whereas $\sigma_L(kz)$ which as usual, is defined by (18), can have a pole in certain models[52]. When it does have a pole all the conclusions of equations (24–25) fail to apply.

If we treat the current of the positive background as unimportant and can divide electronic forces into Coulomb parts and a remainder we can relate $\varepsilon_0 k_s^2$ to the free electron compressibility in the absence of the Coulomb forces, that is

$$\varepsilon_0 k_s^2 \approx ne^2 \left(\frac{\partial n}{\partial p} \right)_T. \tag{40}$$

If the remaining forces are unimportant we can set $\varepsilon_0 = 1$ and obtain $k_s^2 = ne^2\beta$ for classical systems and

$$k_s^2 = \frac{3nme^2}{\hbar^2 k_F^2} = \frac{3nme^2}{\hbar^2} \left(\frac{6\pi^2 n}{2S + 1} \right)^{-2/3}$$

for quantum fermi systems. Correspondingly we obtain

$$u^2 = \frac{\omega_p^2}{k_s^2} \approx \frac{1}{m} \left(\frac{dp}{dn} \right)_T. \tag{41}$$

We shall illustrate these arguments in considerable detail in Sec. G.

4. Transverse Response

In Equation (19) we related the transverse conductivity to the current correlation function. The relation is rather complicated. A far better, more physical and mathematically equivalent way of understanding the transverse conductivity is to relate it to the fields which give rise to these currents. On the basis of relations we shall prove below and the macroscopic Maxwell equations we can, in a straightforward manner, show that equation (19) is precisely equivalent to

$$\varepsilon_T(kz) = 1 + \frac{i\sigma_T(kz)}{z} = \frac{c^2 k^2}{z^2} + \frac{(z^2 - c^2 k^2)^2}{z^2(z^2 - c^2 k^2 + \omega_p^2 - \chi_{JJ}^T(kz))}$$

$$= \frac{c^2 k^2}{z^2} + \left(1 - \chi_{EE}^T(kz) \right)^{-1}$$

$$= \frac{c^2 k^2}{z^2} \left(1 - \chi_{BB}^{-1}(kz) \right) \tag{42}$$

where $\chi_{EE}^T(kz)$ and $\chi_{BB}^T(kz)$ are the usual Hilbert transforms for the transverse electric field and the magnetic field. The last form, i.e.,

$$\chi_{BB}(kz) = c^2 k^2 \left[-z^2 + c^2 k^2 - iz\sigma_T(kz)\right]^{-1} \tag{43}$$

and the next to last are just the type we have encountered so often, in which the transport coefficient occurs in the expression for the inverse correlation function, i.e., it occurs in the differential equation whose solution is the correlation function.

Indeed, using the Maxwell equation

$$\frac{\partial \mathbf{B}}{\partial t} = -c\nabla \times \mathbf{E} \tag{44}$$

and the field commutation relations which lead to (42), we also obtain $\chi_{EE}^T(k0) = 1$ (the transverse analog for arbitrary k of equation (21) $\lim_{k \to 0} \chi_{EE}^L(k0) = 1$). The second form of (42) is therefore precisely our old friend, the dispersion relation of Secs. B, C and D,

$$1 - \frac{\chi_{EE}^T(kz)}{\chi_{EE}^T(k0)} = \frac{z^2}{z^2 - c^2 k^2 + iz\sigma_T(kz)} \tag{45}$$

(where σ_T like Γ in Sec. B can contain a singular term).

If we begin with the magnetic field correlation function, we may proceed as follows. By our usual arguments $\chi_{BB}''(k\omega)$ is real, positive and even. Furthermore,

$$\int \frac{d\omega}{\pi} \omega \chi_{BB}''(k\omega) \left(\delta_{ij} - \frac{k_i k_j}{k^2}\right) = \int d\mathbf{r}\, e^{-i\mathbf{k}\cdot(\mathbf{r}-\mathbf{r}')} \langle i\left[-c(\nabla \times E)_i (\mathbf{r}0), B_j(\mathbf{r}'0)\right]\rangle \tag{46}$$

$$= c\varepsilon_{ikl} k_k \int d\mathbf{r} e^{-i\mathbf{k}\cdot(\mathbf{r}-\mathbf{r}')} \langle [E_l(\mathbf{r}0), B_j(\mathbf{r}'0)]\rangle \tag{47}$$

where we have introduced a summation convention for repeated indices. Using the field commutation relations we may simplify the right hand side of (47) to

$$= -c^2 \varepsilon_{ikl} k_k k_n \varepsilon_{njl} = -c^2 (k_i k_j - k^2 \delta_{ij})$$

and deduce

$$\int \frac{d\omega}{\pi} \omega \chi_{BB}''(k\omega) = c^2 k^2. \tag{48}$$

It follows that for large z

$$\chi_{BB}(kz) = \int \frac{d\omega}{\pi} \frac{\omega \chi_{BB}''(k\omega)}{\omega^2 - z^2} \to -\frac{1}{z^2} c^2 k^2 \tag{49}$$

and that we may write for χ_{BB}^{-1}, which is analytic,

$$\frac{c^2k^2}{z^2\chi_{BB}(kz)} - 1 + \frac{c^2k^2}{z^2\chi_{BB}(k0)} \equiv \int \frac{d\omega}{\pi} \frac{\sigma_T'^{\text{reg}}(k\omega)}{\omega^2 - z^2}$$

$$\equiv i\sigma_T^{\text{reg}}(kz)/z \equiv \varepsilon_T^{\text{reg}}(kz) - 1 \qquad (50)$$

where $\sigma_T^{\text{reg}}(k\omega)$ is a real even function which contains no term in $\delta(\omega)$. Comparison with (19) shows that

$$\sigma_T'(k\omega) = \sigma_T'^{\text{reg}}(k\omega) + c^2k^2 [\chi_{BB}^{-1}(k0) - 1] \pi\delta(\omega) \qquad (51)$$

$$\sigma_T(kz) = \sigma_T'^{\text{reg}}(kz) + ic^2k^2 [\chi_{BB}^{-1}(k0) - 1]/z. \qquad (52)$$

From (44) and (45) we also have the relations

$$c^2k^2\chi_{EE}^T(kz) = z^2\chi_{BB}(kz) + c^2k^2 \qquad (53)$$

$$\chi_{EE}^T(k0) = 1 \qquad (54)$$

which show the equivalence of the last two definitions in Eq. (42) and the connection between (50) and the dispersion relation of Sec. B for $\chi_{EE}^T(kz)$. Finally, let us note that the other microscopic Maxwell equation,

$$c\nabla \times \mathbf{B} = \dot{\mathbf{E}} + \mathbf{J}; -c^2\nabla^2\mathbf{B} + \ddot{\mathbf{B}} = c\nabla \times \mathbf{J} \qquad (55)$$

gives

$$(\omega^2 - c^2k^2)^2 \chi_{BB}''(k\omega) = c^2k^2\chi_{JJ}''^T(k\omega) \qquad (56)$$

which yields the remaining equality in (42).

These equations are therefore our standardized equations, except that we now use the notation

$$c^2k^2\chi_{BB}^{-1}(k0) \equiv c^2k^2 \mu^{-1}(k) \quad \text{for } \omega_0^2 \text{ or } c_0^2(k)k^2$$

and

$$\sigma_T^{\text{reg}}(kz) \quad \text{for } \gamma(z) \quad \text{or} \quad D(kz)k^2. \qquad (57)$$

The standard moment sum rule for $\chi_{EE}''^T(k\omega)$

$$\int_{-\infty}^{\infty} \frac{d\omega}{\pi} \omega^3\chi_{B_iB_j}(k\omega) = \int d\mathbf{r} \, e^{-i\mathbf{k}\cdot\mathbf{r}} \left\langle \left[\frac{i}{\hbar} \ddot{\mathbf{B}}_i(\mathbf{r}0), \dot{\mathbf{B}}_j(0) \right] \right\rangle$$

$$= \int_{-\infty}^{\infty} \frac{d\omega}{\pi} \omega\chi_{E_iE_j}''^T(k\omega) \, c^2k^2$$

$$= c^2k^2(c^2k^2 + \omega_p^2) (\delta_{ij} - k_ik_j/k^2) \qquad (58)$$

9

now gives for the analog of (B. 52)

$$\int_{-\infty}^{\infty} \frac{d\omega}{\pi} \sigma_T'(k\omega) = \omega_p^2$$

$$\int_{-\infty}^{\infty} \frac{d\omega}{\pi} \sigma_T'^{\text{reg}}(k\omega) = \omega_p^2 - c^2 k^2 \left(\mu^{-1}(k) - 1 \right). \tag{59}$$

This proof of the transverse sum rule and Kramers-Kronig relation, which involves the dynamics of the transverse electromagnetic field, is the only rigorous one I know.[57] When $\mu^{-1}(k)$ is not singular, we have, as $k \to 0$

$$\sigma_T'^{\text{reg}}(0\omega) = \sigma_T'(0\omega). \tag{60}$$

Both are also equal in this limit to $\sigma_L(0\omega) = \bar{\sigma}_L(0\omega)$. We also have the "thermodynamic sum rule"

$$\lim_{k \to 0} \chi_{BB}(k0) \equiv \lim_{k \to 0} \mu(k) \equiv \mu = \frac{\partial B}{\partial H} = \lim_{k \to 0} \int \frac{d\omega}{\pi} \frac{\chi_{BB}''(k\omega)}{\omega} \tag{61}$$

for small k, where the dispersion equation

$$\chi_{BB}(kz) = [c^2 k^2 \mu^{-1}(k) - z^2 \varepsilon_T^{\text{reg}}(kz)]^{-1} c^2 k^2$$

$$= [c^2 k^2 \mu^{-1}(k) - z^2 - iz \sigma_T^{\text{reg}}(kz)]^{-1} c^2 k^2 \tag{62}$$

corresponds to the macroscopic statement

$$\nabla \times \nabla \times \mathbf{B} = \nabla \times \left[\frac{1}{c} \mathbf{J}^{\text{free}} + \frac{1}{c} \dot{\mathbf{E}} + \frac{1}{c} \dot{\mathbf{P}} + \nabla \times \mathbf{M} \right]$$

$$- \nabla^2 \mathbf{B} = \nabla \times \frac{1}{c} \sigma' \mathbf{E} + \nabla \times \left(\frac{-i\omega\varepsilon'}{c} \right) \mathbf{E} + \nabla \times \nabla \times \left(\frac{\mu - 1}{\mu} \right) \mathbf{B}$$

$$0 = c^2 k^2 \frac{1}{\mu} \mathbf{B} - \omega^2 (\varepsilon' + i\varepsilon'')^{\text{reg}} \mathbf{B}$$

$$= c^2 k^2 \frac{1}{\mu} \mathbf{B} - \omega^2 \mathbf{B} - i\omega(\sigma' + i\sigma'')^{\text{reg}} \mathbf{B}. \tag{63}$$

In this example we again see why we associated the simple single collision time approximation with Drude's name, i.e., the standard interpolation formula is just

$$\sigma_T^{\text{reg}}(0z) = \sigma_T(0z) = \omega_p^2 \frac{\tau}{1 - iz\tau} \; ; \qquad \sigma_T'(0\omega) = \sigma_T'^{\text{reg}}(0\omega) = \frac{\omega_p^2 \tau}{1 + (\omega\tau)^2}.$$

$$\tag{64}$$

Now note what happens if $\mu^{-1}(k)$ is singular and equal for small k to

$$\frac{1}{\mu_0}\left[\frac{k_L^2}{k^2} + 1\right] \equiv \frac{\varrho_s}{\varrho}\, \frac{\omega_p^2}{c^2 k^2} + \mu_0^{-1}. \tag{65}$$

From Eq. (61), we see that for a uniform field $\partial B/\partial H = 0$. We also have from (51)

$$\sigma_T'(0\omega) = \pi\omega_p^2(\varrho_s/\varrho)\,\delta(\omega) + \sigma_T'^{\mathrm{reg}}(0\omega) \tag{66}$$

and from[58] (59)

$$\int \frac{d\omega}{\pi}\, \sigma_T'^{\mathrm{reg}}(0\omega) = \frac{\varrho - \varrho_s}{\varrho}\, \omega_p^2 \equiv \frac{\varrho_n}{\varrho}\, \omega_p^2. \tag{67}$$

Eq. (66) corresponds in a generalized Drude formula like (64)

$$\sigma_T'(0\omega) = \frac{b_1\omega_p^2\tau_1}{1 + (\omega\tau_1)^2} + \frac{b_2\omega_p^2\tau_2}{1 + (\omega\tau_2)^2}\,;\, b_1 + b_2 = 1 \tag{68}$$

to setting $b_1 = \varrho_s/\varrho$ and $\tau_1 \to \infty$. These are just the equations which describe a superconductor in the London limit. The last equation corresponds to saying the collision time τ for the supercurrent is ∞. Just as the term in $\varepsilon_L(k0) - 1 = \varepsilon_0 k_s^2/k^2$ gives screening of charges, the formula $\mu^{-1}(k) - 1 = k_L^2/k^2\mu_0$ gives the Meissner effect. Indeed the magnetic screening length[59] (the London penetration depth) is k_L^{-1} for the same reason that the charge screening length (the Debye length) is k_s^{-1}.

In substances in which the permeability $\mu(k)$ is not equal to zero the sum rule for the transverse conductivity $\sigma^{T\,\mathrm{reg}}$ is violated only to order k^2. The deficiency describes the often small paramagnetic and diamagnetic effects of the medium. It is easy to prove that in a classical system the orbital contribution to $\chi_{JJ}^T(k0)$ is ω_p^2. With (42) this leads to the van Leeuwen theorem – there is no classical orbital diamagnetism. Quantum mechanically, $\chi_{JJ}^T(k0) - \omega_p^2 = 0(k^2)$, and this remainder gives rise to Landau diamagnetism. In addition we have the spin paramagnetism arising from the spin current part of J.

The sum rules for a superconductor

$$\int \frac{d\omega}{\pi}\, \sigma_T'(k\omega) = \omega_p^2\,;\, \lim_{k\to 0}\int \frac{d\omega}{\pi}\, \sigma_T'^{\mathrm{reg}}(k\omega) = \omega_p^2\, \frac{\varrho_n}{\varrho}\,;$$

$$\lim_{k\to 0}\int \frac{d\omega}{\pi}\, \sigma_L'(k\omega) = \omega_p^2\,;\, \lim_{k\to 0}\int \frac{d\omega}{\pi}\, \sigma_L'^{\mathrm{reg}}(k\omega) = \omega_p^2\, \frac{\varrho_n}{\varrho}\,;$$

$$\int \frac{d\omega}{\pi}\, \bar{\sigma}_L'(k\omega) = \frac{\omega_p^2}{\chi_{EE}^L(k0)} \tag{69}$$

are of course reminiscent of those for a superfluid

$$\lim_{k\to 0} \int \frac{d\omega}{\pi} \frac{\chi_{gg}''^{T}(k\omega)}{\omega} = mn_n$$

$$\int \frac{d\omega}{\pi} \frac{\chi_{gg}''^{L}(k\omega)}{\omega} = mn. \qquad (70)$$

Likewise the classical sum rules $\int d\omega \chi_{JJT}''^{class}(k\omega)/\omega = \int d\omega \sigma_T'^{class}(k\omega) = \pi\omega_p^2$ and $\int d\omega \chi_{ggT}''^{class}(k\omega)/\omega = \pi mn$ give the corresponding statements: there is no orbital diamagnetism, and the moment of inertia of a fluid is always the rigid moment.[60]

Finally observe explicitly that the equations we have derived show that (contrary to the frequently asserted theorem) $\sigma'^{T} = \sigma'^{L}$ is not given rigorously by either of the equal and vanishing expressions[61]

$$\lim_{k\to 0} \frac{\chi_{JJ}''^{T}(k\omega)}{\omega} = \lim_{k\to 0} \frac{\chi_{JJ}''^{L}(k\omega)}{\omega}.$$

Rigorously the correct statement is that under many circumstances (specifically when magnetic effects are small) the expressions which are obtained for $\chi_{JJ}''^{T}(k\omega) = \chi_{JJ}''^{L}(k\omega)$ neglecting the average effect of the transverse electromagnetic field and which may be called $\chi_{JJ}''^{scL}(k\omega)$ and $\chi_{JJ}''^{scT}(k\omega)$ approximate the expressions for $\sigma_L = \sigma_T$ in the presence of the average electromagnetic interaction, i.e.

$$\sigma_L(kz) \cong \frac{1}{iz}\left(\chi_{JJ}^{scL}(kz) - \chi_{JJ}^{scL}(k0)\right); \quad \sigma_T(kz) = \frac{1}{iz}\left(\chi_{JJ}^{scT}(kz) - \chi_{JJ}^{scT}(k0)\right).$$

$$(71)$$

The usually quoted Kubo expressions for the thermal conductivity and thermoelectric power are incorrect for the same reason, We refer the interested reader to the literature[52] for the corrected formulas in these cases.

G. Two Elementary Classical Examples; Free and Coulomb Gases

Thus far we have been concerned with the information equilibrium correlation functions contain, and how this information enables us to describe various phenomena. Such a discussion is vacuous from a theoretical point of view, if we do not know the functions. In the other courses you will hear about systematic methods for determining them. In this section we shall indicate what they are like in the simplest systems—the perfect Boltzmann gas, and the classical Coulomb gas. In deriving the correlation functions for the Coulomb gas, we shall not dwell in any detail on justifying our approximations which are, in fact, rigorous in the low density limit.

1. Perfect Boltzmann Gas

Classically, if is useful to work with the operator

$$f(\mathbf{r}\mathbf{p}t) = \sum_{\alpha} \delta(\mathbf{r} - \mathbf{r}_{\alpha}(t)) \, \delta(\mathbf{p} - \mathbf{p}_{\alpha}(t)) \tag{1}$$

whose expectation value in the presence of a time varying external potential $U(\mathbf{r}t)$, satisfies the equation

$$\frac{\partial \langle f(\mathbf{r}\mathbf{p}t) \rangle}{\partial t} = -\frac{\mathbf{p}}{m} \cdot \nabla \langle f(\mathbf{r}\mathbf{p}t) \rangle + \nabla U(\mathbf{r}t) \cdot \nabla_{\mathbf{p}} \langle f(\mathbf{r}\mathbf{p}t) \rangle. \tag{2}$$

The change in $\langle f(\mathbf{r}\mathbf{p}t) \rangle$ from equilibrium due to an impulsive force

$$U(\mathbf{r}t) = \delta(\mathbf{r} - \mathbf{r}') \, \delta(t - t') \, \delta U \tag{3}$$

is therefore given by

$$\frac{\partial}{\partial t} \delta \langle f(\mathbf{r}\mathbf{p}t) \rangle + \frac{\mathbf{p}}{m} \cdot \nabla \delta \langle f(\mathbf{r}\mathbf{p}t) \rangle = \nabla \delta(\mathbf{r} - \mathbf{r}') \cdot \nabla_{\mathbf{p}} \langle f^0(\mathbf{r}\mathbf{p}t) \rangle \, \delta(t - t') \, \delta U. \tag{4}$$

On the other hand, for this disturbance, we have

$$\delta n(\mathbf{r}t) = \int \delta \langle f(\mathbf{r}\mathbf{p}t) \rangle \, d\mathbf{p} = -\tilde{\chi}_{nn}(\mathbf{r}\mathbf{r}'; t - t') \, \delta U. \tag{5}$$

We may therefore calculate χ from the solution to Eq. (4). By introducing Fourier transforms we easily obtain its solution

$$\frac{\delta n(\mathbf{r}t)}{\delta U} = -\tilde{\chi}_{nn}(\mathbf{r}\mathbf{r}'; t - t')$$

$$= \lim_{\varepsilon \to 0} \int \frac{d\mathbf{k}}{(2\pi)^3} \int \frac{d\omega}{2\pi} \int d\mathbf{p} \, \frac{\mathbf{k} \cdot \nabla_{\mathbf{p}} f^0}{\mathbf{p} \cdot \mathbf{k}/m - (\omega + i\varepsilon)} \, e^{i\mathbf{k} \cdot (\mathbf{r} - \mathbf{r}') - i\omega(t - t')}. \tag{6}$$

The susceptibility is therefore given by

$$\chi_{nn}(kz) = -\int d\mathbf{p} \, \frac{\mathbf{k} \cdot \nabla_{\mathbf{p}} f^0}{\mathbf{p} \cdot \mathbf{k}/m - z} \, ; \tag{7}$$

its absorptive part by

$$\chi''_{nn}(k\omega) = -\pi \int d\mathbf{p} \, (\mathbf{k} \cdot \nabla_{\mathbf{p}} f^0) \, \delta\!\left(\omega - \frac{\mathbf{k} \cdot \mathbf{p}}{m} \right). \tag{8}$$

In the equilibrium ensemble, f^0 is

$$f^0 = n \left(\frac{\beta}{2\pi m} \right)^{\frac{3}{2}} \exp\!\left(-\frac{\beta p^2}{2m} \right) \tag{9}$$

(the constant being determined by the condition $\int f^0 dp = n$) so that the absorptive part of the susceptibility is

$$\chi_{nn}''(k\omega) = \frac{1}{2} n\beta \sqrt{2\pi} \left(\frac{\omega}{kv}\right) \exp\left[-\frac{1}{2}\left(\frac{\omega}{kv}\right)^2\right] \tag{10}$$

where v, the thermal velocity, is defined by $\beta(mv^2) = 1$. A plot of the absorption, Fig. 27a shows that it dies off rapidly above kv.

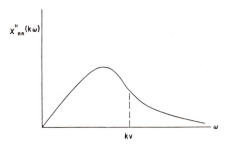

Fig. 27a. The absorptive part of the density response function for a free gas.

Perhaps the easiest way of understanding this is quantum mechanical. Consider a particle in the medium with momentum mv. If it changes its momentum by an amount $m\delta v = \hbar k$, its energy is changed by

$$\delta E = mv \cdot \delta v + \frac{m(\delta v)^2}{2} \cong mv \, \delta v = \hbar k \cdot v$$

$$= \hbar \omega. \tag{11}$$

Thus the distribution in ω/k is essentially the same as the velocity distribution. Note further that χ'' can be explicitly shown to satisfy the sum rules[62]

$$\int \frac{d\omega}{\pi} \chi_{nn}''(k\omega) \, \omega = \frac{nk^2}{m} \; ; \int \frac{d\omega}{\pi} \chi_{nn}''(k\omega) \, \omega^3 = \frac{nk^4}{m^2} \frac{2\langle K.E.\rangle}{\langle N\rangle} = \frac{3nk^4}{m^2\beta} \tag{12}$$

and (since $p = n/\beta$)

$$\lim_{k\to 0} \chi_{nn}(k0) = \lim_{k\to 0} \int \frac{d\omega}{\pi} \frac{\chi_{nn}''(k\omega)}{\omega} = n\beta = n \left.\frac{\partial n}{\partial p}\right|_\beta. \tag{13}$$

The dispersive response, χ', is given by the incomplete error function

$$\chi'(k\omega) = \frac{n\beta}{\sqrt{2\pi}} P \int d\omega' \frac{\omega' \exp\left[-\frac{1}{2}(\omega'/kv)^2\right]}{kv(\omega' - \omega)} = \frac{n}{mv^2} P \int \frac{dx}{\sqrt{2\pi}} \frac{xe^{-\frac{1}{2}x^2}}{x - (\omega/kv)}. \tag{14}$$

This function is depicted in Fig. 27b and because of its occurrence in problems of this kind, it has been extensively tabulated.

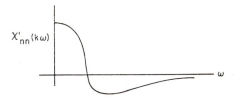

Fig. 27b. The reactive part of the density response function for a free gas.

At vanishing frequency, independent of k, the value of χ' is

$$\chi'_{nn}(k0) = n\beta; \tag{15}$$

its spatial Fourier transform is

$$\chi'_{nn}(\mathbf{r}0) = n\beta\delta(\mathbf{r}). \tag{16}$$

The correlation function is determined from χ by the classical fluctuation dissipation theorem which states

$$S_{nn}(k\omega) = \frac{2}{\beta\omega}\chi''_{nn}(k\omega). \tag{17}$$

Thus the function

$$S_{nn}(k\omega) = \frac{n}{kv}\sqrt{2\pi}\exp\left[-\frac{1}{2}\left(\frac{\omega}{kv}\right)^2\right] \tag{18}$$

observed in neutron scattering experiments, will appear as in Fig. 28. The static correlation function may be immediately written down since

$$\tilde{S}(k0) = \int \frac{d\omega}{2\pi}S(k\omega) = \int \frac{d\omega}{2\pi}\frac{2\chi''(\omega)}{\beta\omega} = \frac{1}{\beta}\chi'(k0). \tag{19}$$

From (15) and (16) we therefore have the instantaneous correlation functions

$$\tilde{S}_{nn}(k0) = n; \quad \tilde{S}_{nn}(\mathbf{r}0) = n\delta(\mathbf{r}). \tag{20}$$

The time dependence of the correlation function is determined by inversion of the Fourier transform of $S(k\omega)$. Thus the time dependence of the correlation function is given by

$$\tilde{S}_{nn}(kt) = n\int_{-\infty}^{\infty}\frac{dx}{\sqrt{2\pi}}e^{-\frac{1}{2}x^2}e^{-ikv\,xt} = n\exp\left[-\frac{1}{2}k^2v^2t^2\right] \tag{21}$$

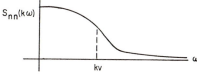

Fig. 28. The density correlation function for the free gas.

and the space-time dependence by the fourier transform of this Gaussian

$$\tilde{S}_{nn}(\mathbf{r}t) = \frac{n \exp\left[-\frac{1}{2}(r^2/v^2 t^2)\right]}{2(2\pi)^{3/2}(vt)^3}. \tag{22}$$

Since there are no collisions, there is no hydrodynamic region and the formulae for the transport coefficients are inapplicable.

2. Coulomb Gas, Approximate Susceptibility

In order to discuss correlations in interacting classical systems systematically it is necessary to discuss the coupled Liouville equations and their rearrangement. While this is feasible, it is an extensive program. Therefore, we shall discuss only the simplest interesting interacting system in the lowest approximation. That system, in which the collisions can still be neglected, is the classical electron gas in a uniform background of positive charge. The point is that at low densities, because of the long range of the Coulomb force, *the interparticle force* can be viewed as an average potential in which the particles move. The effect of this average potential is to induce a charge, which may be calculated in perturbation theory in terms of the non-interacting system.

Thus the approximate polarizability of the non-interacting system, i.e., the function, $\varepsilon_L - 1$, introduced in the last chapter is the longitudinal current correlation for a non-interacting system

$$\frac{i\sigma_L(kz)}{z} = \frac{1}{z^2}\left[\chi_{JJ}^{0L}(kz) - \frac{ne^2}{m}\right] = \frac{1}{k^2}\chi_{\varrho\varrho}^0(kz) = \frac{e^2}{k^2}\chi_{nn}^0(kz). \tag{23}$$

It follows from the rigorous relation

$$\frac{i\sigma_L}{z} = 1 - \left[1 - \frac{e^2}{k^2}\chi_{nn}(kz)\right]^{-1} \tag{24}$$

that

$$1 + \frac{e^2}{k^2}\chi_{nn}^0(kz) \cong \left[1 - \frac{e^2}{k^2}\chi_{nn}(kz)\right]^{-1}. \tag{25}$$

We can make this same statement in a more precise but less general way by looking at the Liouville equation

$$\frac{\partial}{\partial t}\langle f_1\rangle + \frac{\mathbf{p}}{m}\cdot\nabla\langle f_1\rangle - \int \nabla v(\mathbf{r}-\mathbf{r}')\cdot\nabla_{\mathbf{p}}\langle f_2\rangle d\mathbf{r}'d\mathbf{p}' = e\nabla\,\varphi^{\text{ext}}\cdot\nabla_{\mathbf{p}}\langle f_1\rangle \tag{26}$$

where

$$\langle f_2\rangle = \langle\sum_{\alpha\neq\beta}\delta(\mathbf{r}-\mathbf{r}_\alpha(t))\,\delta(\mathbf{p}-\mathbf{p}_\alpha(t))\,\delta(\mathbf{r}'-\mathbf{r}_\beta(t))\,\delta(\mathbf{p}'-\mathbf{p}_\beta(t))\rangle \tag{27}$$

and making the approximation that the particles are uncorrelated, that is, that $f_2(\mathbf{r}\mathbf{p}t;\ \mathbf{r}'\mathbf{p}'t) \cong f_1(\mathbf{r}\mathbf{p}t) f_1(\mathbf{r}'\mathbf{p}'t)$. In the presence of the disturbance $\varphi_{ext}(\mathbf{r}t) = \delta\varphi_{ext}\,\delta(\mathbf{r} - \mathbf{r}')\,\delta(t - t')$, we solve for the linear deviation δf in the distribution function

$$f_1 = f_1^0 + \delta f. \tag{28}$$

The linear response satisfies the equation

$$\frac{\partial}{\partial t}\delta\langle f(\mathbf{r}\mathbf{p}t)\rangle + \frac{\mathbf{p}}{m}\cdot\nabla\delta\langle f(\mathbf{r}\mathbf{p}t)\rangle - \int \nabla v(\mathbf{r} - \bar{\mathbf{r}})\cdot\nabla_{\mathbf{p}}\left[f_0(p)\,\delta\langle f(\overline{\mathbf{r}\mathbf{p}}t)\rangle\right.$$

$$+ f_0(\bar{p})\,\delta\langle f(\mathbf{r}\mathbf{p}t)\rangle]\,d\bar{\mathbf{r}}\,d\bar{\mathbf{p}} = e\,\nabla\delta(\mathbf{r} - \mathbf{r}')\,\delta(t - t')\cdot\nabla_{\mathbf{p}}f_0. \tag{29}$$

Furthermore, as in (6)

$$e \int d\mathbf{p}\,\delta f(\mathbf{r}\mathbf{p}t) = -\bar{\chi}_{\varrho\varrho}(\mathbf{r}\mathbf{r}';t - t'). \tag{30}$$

The second term in the potential integral in (29) vanishes since

$$\int d\bar{\mathbf{r}}\nabla v(\mathbf{r} - \bar{\mathbf{r}}) = - \int d\bar{\mathbf{r}}\bar{\nabla}v(\bar{\mathbf{r}}) = 0. \tag{31}$$

Therefore, when we introduce Fourier transforms we obtain

$$\left(-\omega + \frac{\mathbf{p}}{m}\cdot\mathbf{k}\right)\delta f(\mathbf{k}\mathbf{p}\omega) - v(k)\,\mathbf{k}\cdot\nabla_{\mathbf{p}}f_0(p)\int\delta f(\mathbf{k}\bar{\mathbf{p}}\omega)\,d\bar{\mathbf{p}} = e\mathbf{k}\cdot\nabla_{\mathbf{p}}f_0(p)\,\delta\varphi^{ext},$$
$$\tag{32}$$

or since

$$\bar{\chi}_{\varrho\varrho}(\mathbf{r}\mathbf{r}';t - t') = \int \frac{d\omega}{2\pi}\int\frac{d\mathbf{k}}{(2\pi)^3}\chi_{\varrho\varrho}(k\omega)\,e^{-i\omega(t-t') + i\mathbf{k}\cdot(\mathbf{r}-\mathbf{r}')}$$

$$\chi_{\varrho\varrho}(k\omega) = -e^2 \int d\mathbf{p}\,\delta f(\mathbf{k}\mathbf{p}\omega)$$

$$= -\left[1 - v(k)\int\frac{\mathbf{k}\cdot\nabla_{\mathbf{p}}f_0(p)\,d\mathbf{p}}{\mathbf{p}\cdot\mathbf{k}/m - \omega}\right]^{-1}\int\frac{e^2\mathbf{k}\cdot\nabla_{\mathbf{p}}f_0(p)d\mathbf{p}}{\mathbf{p}\cdot\mathbf{k}/m - \omega}. \tag{33}$$

The function $\chi_{\varrho\varrho}(k\omega)$ is therefore

$$\chi_{\varrho\varrho}(k\omega) = \frac{e^2\chi_{nn}^0(k\omega)}{1 + v(k)\,\chi_{nn}^0(k\omega)} = \frac{1}{v(k)}\left[1 - \frac{1}{1 + v(k)\,\chi_{nn}^0(k\omega)}\right]. \tag{34}$$

This kind of time dependent self-consistent field approximation is sometimes called a random phase approximation. It also applies for non-Coulombic forces when collisions are not important, but the average effect of other particles is important and can be treated as an effective field. Then it gives rise to a phenomenon called zero sound in quantum systems. Eq. (34) is a special case of the formula we discussed earlier (e.g. in (23)) since if we let $\chi_{\varrho\varrho}^{sc}(kz) = \chi_{nn}^0(kz)\, e^2$

$$[1 - v(k)\, \chi_{nn}(kz)]^{-1} = 1 + i\sigma_L(kz)/z$$

$$\cong 1 + v(k)\, \chi_{nn}^0(kz). \tag{35}$$

Note that in this approximation the additional sum rules (24) and (25) are satisfied and, in particular that

$$\lim_{k\to 0} \int \frac{d\omega}{\pi}\, \frac{k^2}{e^2}\, \frac{\sigma_L'(k\omega)}{\omega} = \lim_{k\to 0} \int \frac{d\omega}{\pi}\, \frac{\chi_{nn}''(k\omega)}{\omega}\, |\varepsilon_L|^2 \equiv \lim_{k\to 0} \int \frac{d\omega}{\pi}\, \frac{\chi_{nn}''^0(k\omega)}{\omega}$$

$$= n\left(\frac{\partial n}{\partial p}\right)_T \tag{36}$$

where $n(\partial n/\partial p)_T$ is here the compressibility of the free carriers without the compensating background charge and their own Coulomb charge.

Let us see how the approximation (35) leads to all the results for the thermodynamics and correlations in the Coulomb gas. Since we have not developed the necessary formalism for analyzing the error we will merely state without proof that the above approximation gives rigorously the dominant corrections when the parameter which characterizes the system, $\beta e^2 n^{1/3}$, is much less than unity.[63] Basically, the important property is that the potential energy $e^2 n^{1/3}$ be much less than the kinetic energy $1/\beta$. (Quantum mechanically the same approximations[64] apply when $e^2 n^{1/3}$ is less than the fermi energy $\hbar^2 n^{2/3}/2m$, or equivalently, at high densities, when the interparticle separation is much smaller than a Bohr radius.) We will first discuss the correlation functions and then the thermodynamics. Let us examine the response function

$$\frac{e^2}{k^2}\, \chi_{nn}(k\omega) = 1 - \left[1 + \frac{e^2}{k^2}\, \chi_{nn}^0(k\omega)\right]^{-1} = 1 - \varepsilon_L^{-1}(k\omega). \tag{37}$$

Note first from (15) that when $\omega = 0$, $\chi_{nn}^0(k0) = n\beta$ for all k. Thus $\varepsilon_L(k0) = 1 + i\sigma_L/z = 1 + ne^2\beta/k^2$, and so the potential due to a stationary charge in the system is

$$\varphi(k) = \frac{1}{k\varepsilon_L^2(k0)} = \frac{1}{k^2 + ne^2\beta} \tag{38}$$

The charge is screened, and the screening radius is $(ne^2\beta)^{-1/2}$. Of course in space this gives rise to a potential

$$\varphi(r) = \frac{e^{-k_s r}}{4\pi r} \tag{39}$$

where $k_s^2 = ne^2\beta$.

The absorptive response is given by

$$\chi''_{nn}(k\omega) = \frac{\chi^{0''}_{nn}(k\omega)}{\left[1 + \frac{e^2}{k^2}\chi^{0'}_{nn}(k\omega)\right]^2 + \frac{e^4}{k^4}[\chi^{0''}_{nn}]^2}. \tag{40}$$

For small wave numbers the first term in the denominator can be written as

$$1 + k_s^2 P \int_0^\infty \frac{dx}{\sqrt{2\pi}} e^{-\frac{1}{2}x^2} \frac{2x^2 v^2}{v^2 x^2 k^2 - \omega^2}$$

$$\to 1 - \frac{k_s^2 v^2}{\omega^2}\left(\int_0^\infty \frac{dx}{\sqrt{2\pi}} e^{-\frac{1}{2}x^2} 2x^2\left[1 + \frac{x^2 k^2 v^2}{\omega^2}\right]\right) \tag{41}$$

$$\to 1 - \frac{ne^2}{m\omega^2}\left[1 + \left(\frac{k}{k_s}\right)^2 \frac{3ne^2}{m\omega^2}\right]. \tag{42}$$

This expression vanishes when

$$\omega^2 \cong \omega_p^2\left[1 + \frac{3k^2}{k_s^2}\right] \tag{43}$$

where $\omega_p^2 \equiv ne^2/m$ is the plasma frequency. For small values of k we have

$$\frac{e^2}{k^2}\chi^{0''}_{nn}(k\omega) \cong \frac{k_s^2}{k^2}\left(\frac{\pi}{2}\right)^{\frac{1}{2}} \frac{\omega}{kv} \exp\left[-\frac{1}{2}\left(\frac{\omega}{kv}\right)^2\right] \tag{44}$$

which is extremely small for $\omega/kv >> 1$. Thus in the neighborhood of the plasma frequency $\omega^2 \sim \omega_p^2$, the damping is extremely small. In this region χ'' may be approximated as

$$\chi''_{\varrho\varrho}(k\omega) \approx \frac{nk^2 e^2}{m} \mathrm{Im}\left[\omega^2 - \omega_p^2\left(1 + \frac{3k^2}{k_s^2}\right) - i\omega^2\left(\frac{\pi}{2}\right)^{\frac{1}{2}}\left(\frac{k_s}{k}\right)^3 \exp\left(-\frac{k_s^2}{2k^2}\right)\right]^{-1} \tag{45}$$

so that there are well defined sharp plasma modes for which

$$\omega \approx \omega_p\left[\pm\left(1 + \frac{3k^2}{k_s^2}\right) - i\left(\frac{\pi}{8}\right)^{\frac{1}{2}}\left(\frac{k_s}{k}\right)^3 \exp\left(-\frac{k_s^2}{2k^2}\right)\right]. \tag{46}$$

The small damping of these very sharp resonances is called Landau damping. The fact that

$$\int \frac{d\omega}{\pi} \omega \frac{nk^2}{m} \pi\delta(\omega^2 - \omega_p^2) = \frac{nk^2}{m} \tag{47}$$

indicates that to order k^2, the plasma pole exhausts the sum rule.

A plot of $\chi''(k\omega)$ for small k appears in Fig. 29. For values of k which are much larger than k_s, on the other hand, the denominator in the function χ is given by

$$1 + \frac{k_s^2}{k^2} \int_{-\infty}^{\infty} \frac{dx}{\sqrt{2\pi}} \frac{x^2 e^{-\frac{1}{2}x^2}}{x^2 - (\omega/kv)^2} \approx 1 \tag{48}$$

and so the Coulomb gas values are like the free gas values.

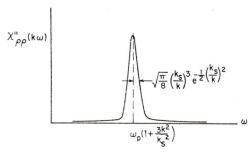

Fig. 29. The absorptive part of the charge density response function for small k. A plasma resonance which is Landau damped dominates the response.

3. Application of $\chi''(k\omega)$ to Energy Loss Calculations

With the function $\chi(k\omega)$ we can treat various problems dealing with the interaction of charges and matter. One which is straightforward but has been particularly badly treated in the literature is the problem of the energy loss of a particle travelling through the medium due to its interaction with the charges in the medium. We may calculate the rate at which a particle loses energy since it is equal to the rate at which it does work on the system. The rate at which it does work on the system by interchanging momentum k and energy $E(\mathbf{p}) - E(\mathbf{p} - \hbar\mathbf{k})$ with the system is given by the matrix element squared for a transition in which the particle changes its momentum and energy by

$$\mathbf{p} \to \mathbf{p} - \hbar\mathbf{k}$$

$$E(\mathbf{p}) \to E(\mathbf{p} - \hbar\mathbf{k})$$

multiplied by the density of states for the particle and the system. By the standard arguments of the second section this expression

$$\frac{W}{T} = \int \sum_{\substack{\text{final} \\ \text{states}}} 2\pi\omega \varrho_{\text{system}} \frac{d\mathbf{k}}{(2\pi)^3} d\omega \delta(\omega - [E(\mathbf{p}) - E(\mathbf{p} - \hbar\mathbf{k})]/\hbar)$$

$$\times \left| \left\langle E_{\text{system}} + \hbar\omega, \, \mathbf{p} - \hbar\mathbf{k} \left| \frac{1}{c} \mathbf{J} \cdot \mathbf{A} \right| E_{\text{system}}, \, \mathbf{p} \right\rangle \right|^2 \tag{49}$$

may be written in the form

$$\frac{W}{T} = \int \int \frac{d\mathbf{k}}{(2\pi)^3} \frac{d\omega}{\omega} \chi''_{E_i E_j}(k\omega) \left(\frac{2\mathbf{p} - \hbar\mathbf{k}}{2M}\right)_i \left(\frac{2\mathbf{p} - \hbar\mathbf{k}}{2M}\right)_j$$

$$\times e^2 \delta(\omega - [E(\mathbf{p}) - E(\mathbf{p} - \hbar\mathbf{k})]/\hbar)$$

$$= \int \frac{d\mathbf{k}}{(2\pi)^3} \chi''_{E_i E_j}\left(k \frac{(2\mathbf{p} - \hbar\mathbf{k}) \cdot \mathbf{k}}{2M}\right) e^2 \left(\frac{2\mathbf{p} - \hbar\mathbf{k}}{2M}\right)_i \left(\frac{2\mathbf{p} - \hbar\mathbf{k}}{2M}\right)_j \frac{2M}{(2\mathbf{p} - \hbar\mathbf{k}) \cdot \mathbf{k}} .$$

$$(50)$$

Because of Maxwell's equations, the field correlation function is related to the current correlation function. While there are contributions from the transverse field correlations, for slowly moving particles $v \ll c$, the longitudinal contributions dominate. Furthermore, since the field satisfies the equation

$$i\mathbf{k} \cdot \mathbf{E} = \varrho = \frac{\mathbf{k} \cdot \mathbf{J}}{\omega}$$

this *longitudinal contribution* may be written as

$$\frac{W}{T} = \int \frac{d\mathbf{k}}{(2\pi)^3} \chi''_{nn}\left(k \frac{(2\mathbf{p} - \hbar\mathbf{k}) \cdot \mathbf{k}}{2M}\right) \frac{e^4}{k^4} \frac{(2\mathbf{p} - \hbar\mathbf{k}) \cdot \mathbf{k}}{2M} \qquad (51)$$

a form which is similar to the form usually derived by considering a classical particle of mass M moving with velocity $\mathbf{v}_0$ and neglecting its recoil,

$$\frac{W}{T} = \int \frac{d\mathbf{k}}{(2\pi)^3} \int \chi''_{nn}(k\omega) \frac{e^4}{k^4} \omega\delta\left(\omega - \frac{\mathbf{k} \cdot \mathbf{p}}{M}\right) d\omega. \qquad (52)$$

One derives (52) by calculating the work done by a charge whose potential is

$$\varphi^{\text{ext}}(\mathbf{r}t) = \frac{e}{4\pi |\mathbf{r} - \mathbf{v}_0 t|} = e \int \frac{d\mathbf{k}}{(2\pi)^3} \frac{e^{i\mathbf{k} \cdot (\mathbf{r} - \mathbf{v}_0 t)}}{k^2} \qquad (53)$$

and substituting into

$$W = \int d\mathbf{r} \int dt \int d\mathbf{r}' \int dt' \, \varphi^{\text{ext}}(\mathbf{r}t) \left[i \frac{\partial}{\partial t} \bar{\chi}''(\mathbf{r}\mathbf{r}'; t - t')\right] \varphi^{\text{ext}}(\mathbf{r}'t'). \quad (54)$$

The difference between (52) and (51) is that the former takes account of the recoil experienced by the particle. The formulas obtained from (51) and (52) are quite complicated but their qualitative features can be easily understood. To do so we first observe that if we neglect recoil and the effects of the coulomb force, we obtain from (52)

$$\int \frac{d\mathbf{k}}{(2\pi)^3} \frac{e^4}{k^4} (\mathbf{k} \cdot \mathbf{v}_0)^2 \exp\left[-\frac{1}{2}\left(\frac{\mathbf{k} \cdot \mathbf{v}_0}{kv}\right)^2\right] \frac{n\beta}{kv} \left(\frac{\pi}{2}\right)^{\frac{1}{2}} \qquad (55)$$

which reduces to

$$= e^4 \int \frac{dk}{4\pi^2 k} \int\limits_{-1}^{1} d\mu \, \mu^2 \exp\left[-\frac{1}{2}\left(\frac{\mu v_0}{v}\right)^2 \right] n\beta \left(\frac{\pi}{2}\right)^{\frac{1}{2}} \frac{v_0^2}{v} \qquad (56)$$

$$= \frac{ne^4}{\pi^2 m v_0} \int\limits_{0}^{v_0/v} dx \, \frac{x^2}{2} \, e^{-\frac{1}{2}x^2} \left(\frac{\pi}{2}\right)^{\frac{1}{2}} \int \frac{dk}{k} . \qquad (57)$$

We of course are only interested in velocities much greater than the thermal velocity so that this reduces to[65]

$$\frac{W}{T} \cong \frac{ne^4}{4\pi m v_0} \log\left(\frac{k_{\max}}{k_{\min}}\right). \qquad (58)$$

The physical cut offs on this integral are twofold. In the first place at small momenta, the coulomb matrix element (e^4/k^4) is reduced due to screening. This screening is present in the function χ'' which occurs in both (51) and (52) but not in the approximation to it, (55), which holds when $k >> k_s$ and recoil can be neglected. Secondly, at large values of the momentum transfer, the recoil of the particle is important. As we see from the effectively rigorous expression (51) (but not (52)), the upper cut-off in

$$\frac{W}{T} = \int \frac{dk}{(2\pi)^3} \, \chi''_{nn}\left(k \, \frac{(2\mathbf{p} - \hbar\mathbf{k}) \cdot \mathbf{k}}{2M}\right) \frac{e^4}{k^4} \, \frac{(2\mathbf{p} - \hbar\mathbf{k}) \cdot \mathbf{k}}{2M} \qquad (59)$$

is determined[66] by the maximum possible momentum transfer (for which $E_0 = \hbar^2 k_{\max}^2/2M$). It corresponds to the inverse distance of closest approach. Both effects are included in (51); no "joining" of different calculations is necessary. The formula for Čerenkov radiation energy loss follows in a similar fashion, the imaginary part of the transverse field correlation function at $\omega^2 + i\omega\sigma_T = c^2 k^2$ producing the familiar radiation cone.[67]

4. Coulomb Gas Correlations and Thermodynamic Properties

We calculate the correlation function by dividing $\chi''_{\varrho\varrho}(k\omega)$ by $\beta\omega/2$

$$S_{\varrho\varrho}(k\omega) = \frac{2}{\beta\omega} \, \chi''_{\varrho\varrho}(k\omega). \qquad (60)$$

The instantaneous correlation function is therefore

$$\tilde{S}_{\varrho\varrho}(k0) = \int \frac{d\omega}{2\pi} \, \frac{2}{\beta\omega} \, \chi''_{\varrho\varrho}(k\omega) = \frac{1}{\beta} \, \chi_{\varrho\varrho}(k0) = \frac{1}{\beta} \, \frac{k_s^2 k^2}{k^2 + k_s^2} , \qquad (61)$$

$$\tilde{S}_{\varrho\varrho}(r0) = \frac{k_s^2}{\beta} \left[-\frac{k_s^2 \exp(-k_s r)}{4\pi r} + \delta(\mathbf{r}) \right]. \qquad (62)$$

Note that since

$$\tilde{S}_{\varrho\varrho}(\mathbf{r} - \mathbf{r}'; 0) = e^2 [\langle \sum_\alpha \delta(\mathbf{r} - \mathbf{r}_\alpha) \sum_\beta \delta(\mathbf{r}' - \mathbf{r}_\beta) \rangle] - \langle \varrho(\mathbf{r}) \rangle \langle \varrho(\mathbf{r}') \rangle$$

we have

$$e^2 \langle \sum_{\alpha \neq \beta} \delta(\mathbf{r} - \mathbf{r}_\alpha) \delta(\mathbf{r}' - \mathbf{r}_\beta) \rangle = \tilde{S}_{\varrho\varrho}(\mathbf{r} - \mathbf{r}'; 0) - e\delta(\mathbf{r} - \mathbf{r}') \langle \varrho(\mathbf{r}) \rangle$$

$$+ \langle \varrho(\mathbf{r}) \rangle \langle \varrho(\mathbf{r}') \rangle. \tag{63}$$

Consequently

$$-\frac{ne^2 k_s^2}{4\pi |\mathbf{r} - \mathbf{r}'|} e^{-k_s|\mathbf{r}-\mathbf{r}'|} = e^2 \langle \sum_{\alpha \neq \beta} \delta(\mathbf{r} - \mathbf{r}_\alpha) \delta(\mathbf{r}' - \mathbf{r}_\beta) \rangle - \langle \varrho(\mathbf{r}) \rangle \langle \varrho(\mathbf{r}') \rangle. \tag{64}$$

After we take into account that the last term cancels with the background charge we obtain for the expectation value of the potential energy

$$V\frac{1}{2}\int d\mathbf{r} \left[-\frac{ne^2}{4\pi r} k_s^2 e^{-k_s r} \frac{1}{4\pi r} \right] = -\frac{1}{2} Ne^2 \int_0^\infty \frac{dr k_s^2 e^{-k_s r}}{4\pi}$$

$$= \frac{-Ne^2(ne^2\beta)^{\frac{1}{2}}}{8\pi}. \tag{65}$$

To calculate the thermodynamic properties from the potential energy is straightforward with the aid of the following trick.[68] We note that with a parameter λ, the free energy is given by

$$E_\lambda - TS_\lambda = -\frac{1}{\beta} Q_\lambda \tag{66}$$

where

$$e^{Q_\lambda} = \mathrm{Tr}\, e^{-\beta H_\lambda}$$

in a canonical ensemble. To calculate Q we observe that if we have a hamiltonian with two body forces and a coupling constant, λ times as large as the actual one, then

$$\frac{dQ_\lambda}{d\lambda} = -\beta \frac{\langle \mathrm{P.E.} \rangle_\lambda}{\lambda} \tag{67}$$

where $\langle \mathrm{P.E.} \rangle_\lambda$ is the potential energy with coupling constant, λ. Integrating we find

$$Q = -\beta \int_0^\lambda \frac{d\lambda'}{\lambda'} \langle \mathrm{P.E.} \rangle_{\lambda'} + Q_{\text{free gas}} \tag{68}$$

or specifically in this problem

$$Q = Q_{\text{free gas}} + \beta \frac{Ne^2(ne^2\beta)^{\frac{1}{2}}}{8\pi} \frac{2}{3}. \tag{69}$$

Since the pressure is given by

$$dE - TdS = -pdV$$
$$= d(Q/\beta)_{\beta\text{constant}} \qquad (70)$$

the equation of state is

$$p = \frac{1}{\beta}\left(\frac{dQ}{dV}\right)_\beta = p_0 - \frac{e^2 n (n e^2 \beta)^{\frac{1}{2}}}{24\pi}$$

$$= \frac{n}{\beta}\left[1 - \frac{(n^{\frac{1}{3}} e^2 \beta)^{\frac{3}{2}}}{24\pi}\right]. \qquad (71)$$

This correction term, originally calculated by Debye and Hückel[69] is the first term in a low density or weak coupling expansion, but, as we see it contains the coupling constant and temperature to fractional powers and therefore cannot be calculated from any term in a virial or coupling constant expansion.

H. The Fundamental Correlation Functions

In the previous section, we evaluated some simple correlation functions. In the case the Coulomb gas, we did so by evaluating the function which played the role of the phenomenological function and which we have in various other contexts denoted as γ or Dk^2. Generally speaking we have argued that we expect functions of this type to be more regular and localized than the correlation functions they determine. However, in a systematic microscopic theory it is not always possible to calculate these functions directly, because we have no closed set of equations for them. Mathematically, this difficulty is connected with the fact that we can only have a closed set of equations if we have a complete (canonical) set of dynamical variables. For this reason the systematic microscopic techniques are generally concerned with the time dependent correlations of these fundamental dynamical (canonical) variables. For correlations of these fundamental fields it is possible to go through the same kind of physical and mathematical arguments we have spelled out, but it is also possible to write closed, albeit impossible to exactly solve, integro-differential equations. As in the other cases, functions analogous to γ, Dk^2 or σ play an essential role. In an interacting quantum mechanical system with two body central spin independent forces the fundamental dynamical field is the quantum field ψ which satisfies the operator Schroedinger equation

$$i\hbar\frac{\partial}{\partial t}\psi(1) + \frac{\hbar^2}{2m}\nabla^2\psi(1) - v(1)\,\psi(1) = \int v(12)\,\psi^\dagger(2)\,\psi(2)\,\psi(1)\,d2 \qquad (1)$$

where the variable index includes spin, space, and time and $v(12) = v(|\mathbf{r}_1 - \mathbf{r}_2|)\,\delta(t_1 - t_2)$. We have written this equation in a form parallel to (A1). It is convenient as in (A4) to discuss it by introducing an external source (i.e. $H^{ext}(t) = \int [\psi^\dagger(\mathbf{r}t)\,\eta(\mathbf{r}t) + \psi(\mathbf{r}t)\,\eta^*(\mathbf{r}t)]\, d\mathbf{r}$) so that we have

$$i\hbar \frac{\partial}{\partial t}\langle\psi\rangle_{n.e.} + \frac{\hbar^2}{2m}\nabla^2\langle\psi\rangle_{n.e.} - v(1)\langle\psi\rangle_{n.e.}$$

$$- \int v(12)\langle\psi^\dagger(2)\,\psi(2)\,\psi(1)\rangle_{n.e.} = \eta(1). \qquad (2)$$

For a boson system (and also for a fermion system when the source[70] is given a slightly different interpretation) we are led to study $\hbar\tilde{\chi}_{\psi\psi^\dagger}(11')$ $\equiv G_R(11') = -\delta\langle\psi(1)\rangle/\delta\eta(1')$, the analog of (A7)

$$i\hbar \frac{\partial}{\partial t} G_R(11') + \frac{\hbar^2}{2m}\nabla^2 G_R(11') - v(1) G_R(11')$$

$$- \int V_R(1\bar{1})\, G_R(\bar{1}1')\, d\bar{1} = -\delta(11') \qquad (3)$$

in which the effective internal potential $V_R(11')$, which plays the role of $iz\gamma$ can be non-local as in (A37). In this case, however, we can do much more since we can give rules for calculating $V_R(11')$. These rules are discussed in other courses at this school.

We shall content ourselves in this lecture with preparing a glossary of parallels. We have instead of (B7) for $\tilde{\chi}'' \equiv \tilde{\varrho}/2\hbar$

$$\tilde{\varrho}(\mathbf{r}st;\mathbf{r}'s't') \equiv \langle\psi(\mathbf{r}st)\,\psi^\dagger(\mathbf{r}'s't') - \varepsilon\psi^\dagger(\mathbf{r}'s't')\,\psi(\mathbf{r}st)\rangle \qquad (4)$$

(where $\varepsilon = 1$ for bosons and -1 for fermions) and its transform

$$\tilde{\varrho}(\mathbf{r}st;\mathbf{r}'s't') = \int \frac{d\omega}{2\pi}\, e^{-i\omega(t-t')}\,\varrho(\mathbf{r}s\mathbf{r}'s';\omega). \qquad (5)$$

The function analogous to $\tilde{\chi}$. $\tilde{G}_R/\hbar$, satisfies

$$\tilde{G}_R(\mathbf{r}st;\mathbf{r}'s't') = i\eta(t-t')\,\tilde{\varrho}(\mathbf{r}st;\mathbf{r}'s't') = \int \frac{d\omega}{2\pi}\, e^{-i\omega(t-t')}\, G_R(\mathbf{r}s\mathbf{r}'s';\omega) \qquad \mathbf{(6)}$$

where

$$G_R(\mathbf{r}s\mathbf{r}'s';\omega) = \lim_{\varepsilon\to 0^+} G(\mathbf{r}s\mathbf{r}'s'; z = \omega + i\varepsilon) \qquad (7)$$

and

$$G(\mathbf{r}s\mathbf{r}'s'; z) \equiv \int \frac{d\omega}{2\pi}\, \frac{\varrho(\mathbf{r}s\mathbf{r}'s';\omega)}{\omega - z}. \qquad (8)$$

In analogy to the functions S_{ij} and S_{ji} we now have two functions

$$\tilde{G}_{\rangle}(\mathbf{r}st;\mathbf{r}'s't') \equiv \langle\psi(\mathbf{r}st)\,\psi^\dagger(\mathbf{r}'s't')\rangle - \langle\psi(\mathbf{r}st)\rangle\langle\psi^\dagger(\mathbf{r}'s't')\rangle$$

$$\equiv \int \frac{d\omega}{2\pi}\, G_{\rangle}(\mathbf{r}s\mathbf{r}'s';\omega)\, e^{-i\omega(t-t')} \qquad (9)$$

and

$$\tilde{G}_<(\mathbf{r}st; \mathbf{r}'s't') \equiv \varepsilon[\langle \psi^\dagger(\mathbf{r}'s't')\, \psi(\mathbf{r}st)\rangle - \langle \psi^\dagger(\mathbf{r}'s't')\rangle\, \langle \psi(\mathbf{r}st)\rangle]$$

$$\equiv \int \frac{d\omega}{2\pi}\, G_<(\mathbf{r}s\mathbf{r}'s'; \omega)\, e^{-i\omega(t-t')}. \tag{10}$$

Like $2\chi''(\mathbf{r}\mathbf{r}'; \omega)$, $\varrho(\mathbf{r}s\mathbf{r}'s'; \omega)$ satisfies sum rules

$$\int \frac{d\omega}{2\pi}\, \varrho(\mathbf{r}s\mathbf{r}'s'; \omega)\, \omega^n = \left\langle \left[\left(i\frac{\partial}{\partial t}\right)^n \psi(\mathbf{r}st), \psi^\dagger(\mathbf{r}'s't)\right]_\varepsilon \right\rangle \tag{11}$$

and in particular

$$\int \frac{d\omega}{2\pi}\, \varrho(\mathbf{r}s\mathbf{r}'s'; \omega) = \delta(\mathbf{r} - \mathbf{r}')\, \delta_{ss'}. \tag{12}$$

When the expectation values are taken in a grand canonical ensemble, i.e. when

$$\langle X \rangle = \text{Tr}\, [e^{-\beta(H-\mu N)} X]\, [\text{Tr}\, e^{-\beta(H-\mu N)}]^{-1} \tag{13}$$

the same fluctuation-dissipation argument we used earlier[71] shows that the two hermitian, positive definite matrices $G_>(\mathbf{r}s\mathbf{r}'s'; \omega)$ and $\varepsilon G_<(\mathbf{r}s\mathbf{r}'s'; \omega)$ are related to each other by

$$G_>(\mathbf{r}s\mathbf{r}'s'; \omega) = \varepsilon\, e^{\beta(\hbar\omega-\mu)} G_<(\mathbf{r}s\mathbf{r}'s'; \omega), \tag{14}$$

and to $\varrho(\mathbf{r}s\mathbf{r}'s'; \omega)$ by

$$G_>(\mathbf{r}s\mathbf{r}'s'; \omega) = \frac{1}{1 - \varepsilon\, e^{-\beta(\hbar\omega-\mu)}}\, \varrho(\mathbf{r}s\mathbf{r}'s'; \omega) \tag{15}$$

$$G_<(\mathbf{r}s\mathbf{r}'s'; \omega) = \frac{1}{\varepsilon\, e^{\beta(\hbar\omega-\mu)} - 1}\, \varrho(\mathbf{r}s\mathbf{r}'s'; \omega). \tag{16}$$

As in classical systems it is possible to express the thermodynamic properties in terms of the fundamental correlation functions. In particular the density is given by

$$N = \sum_s \int d\mathbf{r} \left[\int \frac{d\omega}{2\pi}\, \frac{\varrho(\mathbf{r}s\mathbf{r}s; \omega)}{e^{\beta(\hbar\omega-\mu)} - \varepsilon} + \langle \psi^\dagger(\mathbf{r}s)\rangle\, \langle \psi(\mathbf{r}s)\rangle\right] \tag{17}$$

which in a translationally invariant system becomes

$$n = \sum_s \left[\int \frac{d\mathbf{p}}{(2\pi\hbar)^3} \int \frac{d\omega}{2\pi}\, \frac{\varrho_{ss}(\mathbf{p}\omega)}{e^{\beta(\hbar\omega-\mu)} - \varepsilon} + n_{0s}\right], \tag{18}$$

and the energy density is given by

$$E = \sum_s \int d\mathbf{r} \int d\mathbf{r}'\, \delta(\mathbf{r} - \mathbf{r}') \int \frac{d\omega}{2\pi} \left(\frac{-\hbar^2}{4m}\nabla^2 + \frac{1}{2}\hbar\omega\right) \frac{\varrho(\mathbf{r}s\mathbf{r}'s; \omega)}{e^{\beta(\hbar\omega-\mu)} - \varepsilon}$$

$$+ \sum_s \int d\mathbf{r}\left[\frac{1}{2}\mu\langle \psi^\dagger(\mathbf{r}s)\rangle\, \langle \psi(\mathbf{r}s)\rangle + \frac{\hbar^2}{4m}\nabla\langle \psi^\dagger(\mathbf{r}s)\rangle \cdot \nabla\langle \psi(\mathbf{r}s)\rangle\right] \tag{19}$$

which becomes

$$\varepsilon = \sum_s \left[\int \frac{d\mathbf{p}}{(2\pi\hbar)^3} \int \frac{d\omega}{2\pi} \frac{1}{2} \left(\frac{p^2}{2m} + \hbar\omega \right) \frac{\varrho_{ss}(\mathbf{p}\omega)}{e^{\beta(\hbar\omega - \mu)} - \varepsilon} + \frac{1}{2}\mu n_{0s} \right]. \quad (20)$$

(The terms involving n_0 are usually not present but they can occur in condensed Bose systems.)[72] The Wigner or phase space distribution function,

$$n(\mathbf{p}rst) = \int d\bar{\mathbf{r}} e^{i\mathbf{p}\cdot\bar{\mathbf{r}}/\hbar} \left\langle \psi^\dagger \left(\mathbf{r} + \frac{\bar{\mathbf{r}}}{2} st \right) \psi \left(\mathbf{r} - \frac{\bar{\mathbf{r}}}{2} st \right) \right\rangle \quad (21)$$

reduces in a translationally invariant time independent system to

$$n_s(\mathbf{p}) = \int \frac{d\omega}{2\pi} \frac{\varrho_{ss}(\mathbf{p}\omega)}{e^{\beta(\hbar\omega - \mu)} - \varepsilon} + n_{0s}\delta(\mathbf{p})(2\pi\hbar)^3, \quad (22)$$

a positive definite phase space density.

In classical liquids we saw (in footnote 42) that the form of the momentum correlation function was uniquely determined by requiring the excitations to have infinite lifetime $(D = 0)$. The corresponding approximation here, is the Hartree Fock approximation. In particular, if we have a translationally invariant system,[73] and we suppose that

$$\varrho_{ss'}(\mathbf{p}\omega) = 2\pi\hbar\delta(\hbar\omega - E(\mathbf{p}s))\,\delta_{ss'} \quad (23)$$

to satisfy the monochromatic assumption and

$$\int \frac{d\omega}{2\pi} \varrho_{ss}(\mathbf{p}\omega) = 1, \quad (24)$$

we may determine $E(\mathbf{p}s)$ by the condition

$$\int \frac{d\omega}{2\pi} \hbar\omega\varrho_{ss'}(\mathbf{p}\omega) = E(\mathbf{p}s)\,\delta_{ss'} = \int \left\langle \left[i\hbar\frac{\partial}{\partial t}\psi(rst), \psi^\dagger(\mathbf{r}'\,s't) \right]_s \right\rangle$$
$$\times e^{-i\mathbf{p}\cdot(\mathbf{r}-\mathbf{r}')/\hbar}\, d(\mathbf{r} - \mathbf{r}'). \quad (25)$$

If we allow for a magnetic field coupling to the spins and if we allow the system to be in motion with velocity v and express this motion as an alteration in the Hamiltonian we have

$$i\hbar\frac{\partial\psi(rst)}{\partial t} = -2\gamma Hs\,\psi(rst) + \frac{1}{2m}\left(\frac{\hbar}{i}\nabla + m\mathbf{v} \right)^2 \psi(rst)$$
$$+ \sum_s \int d\bar{\mathbf{r}} v(\mathbf{r} - \bar{\mathbf{r}})\, \psi^\dagger(\bar{\mathbf{r}}\bar{s}t)\,\psi(\bar{\mathbf{r}}\bar{s}t)\,\psi(rst). \quad (26)$$

The commutator then yields

$$E(\mathbf{p}s)\,\delta_{ss'} = \int e^{-i\mathbf{p}\cdot(\mathbf{r}-\mathbf{r}')/\hbar}\, d(\mathbf{r} - \mathbf{r}') \left\{ \frac{1}{2m}\left(\frac{\hbar}{i}\nabla + m\mathbf{v} \right)^2 \delta(\mathbf{r} - \mathbf{r}')\,\delta_{ss'} \right.$$
$$- 2\gamma Hs\,\delta(\mathbf{r} - \mathbf{r}')\,\delta_{ss'} + \int d\bar{\mathbf{r}} \sum_s v(\mathbf{r} - \bar{\mathbf{r}})\langle \psi^\dagger(\bar{\mathbf{r}}\bar{s})\,\psi(\bar{\mathbf{r}}\bar{s})\rangle$$
$$\left. \times \delta(\mathbf{r} - \mathbf{r}')\,\delta_{ss'} + \varepsilon v(\mathbf{r} - \mathbf{r}')\langle \psi^\dagger(\mathbf{r}'s')\,\psi(rs)\rangle \right\}. \quad (27)$$

10*

For spin $\frac{1}{2}$ fermions

$$E(ps) = \frac{(\mathbf{p} + m\mathbf{v})^2}{2m} - 2\gamma s H + \left[\int d\mathbf{r} v(\mathbf{r})\right] \sum_s \int \frac{d\bar{\mathbf{p}}}{(2\pi\hbar)^3} \frac{1}{e^{\beta(E(\bar{p}s) - \mu)} + 1}$$

$$- \int \frac{d\bar{\mathbf{p}}}{(2\pi\hbar)^3} v\left(\frac{\mathbf{p} - \bar{\mathbf{p}}}{\hbar}\right) \frac{1}{e^{\beta(E(\bar{p}s) - \mu)} + 1}. \qquad (28)$$

In analogy with our earlier discussion, if we wish to discuss the correlation function of a fermi system in which $G_{ss'}(\mathbf{p}z) = \delta_{ss'}G_s(\mathbf{p}z)$ it is convenient to consider the quantity

$$G_s(\mathbf{p}z) = \int \frac{d\omega}{2\pi} \frac{\varrho_s(\mathbf{p}\omega)}{\omega - z} \qquad (29)$$

whose inverse exists and is analytic except on the real axis since for fermions $\varrho_s(\mathbf{p}\omega)$, the *sum* of two positive definite matrices, must be greater than zero. It follows that

$$\hbar G^{-1}(\mathbf{p}z) = -\hbar z + E_{\text{H.F.}}(\mathbf{p}) + V(\mathbf{p}z) \qquad (30)$$

where $E_{\text{H.F.}}$ is the energy determined by (27) and

$$\frac{V(\mathbf{p}z)}{\hbar} = -\int \frac{d\omega}{2\pi} \frac{\Gamma'(\mathbf{p}\omega)}{\omega - z} \qquad (31)$$

and where we have omitted the spin label. Equations (29)-(31) also apply to a bose system since the positivity of $(\hbar\omega - \mu)\varrho$ implies that $\text{Im}\,(\hbar z - \mu)\,G(z)$ and hence $G(z) \neq 0$ when $\text{Im}\,z \neq 0$. For qualitative purposes we may think of G as describing the propagation of a bose or fermi excitation through the system. The effect of the medium on the excitation is described by the complex potential V often called the mass operator, and its negative imaginary part $\frac{1}{2}\hbar\Gamma'$ is related to the transition rate. In the Born approximation (neglecting spin factors) this rate, Γ' reduces to

$$\Gamma'(\mathbf{p}\omega) = \frac{2\pi}{\hbar} \int \frac{d\mathbf{p}_2}{(2\pi\hbar)^3} \int \frac{d\mathbf{p}_3}{(2\pi\hbar)^3} \int \frac{d\mathbf{p}_4}{(2\pi\hbar)^3} (2\pi\hbar)^3 \delta(\mathbf{p} + \mathbf{p}_2 - \mathbf{p}_3 - \mathbf{p}_4)$$

$$\times \left(\frac{1}{2}\left|v\left(\frac{\mathbf{p} - \mathbf{p}_3}{\hbar}\right) \pm v\left(\frac{\mathbf{p}_2 - \mathbf{p}_4}{\hbar}\right)\right|^2\right) \delta(\hbar\omega + E(\mathbf{p}_2) - E(\mathbf{p}_3) - E(\mathbf{p}_4))$$

$$\times [(1 + \varepsilon f(\mathbf{p}_3))(1 + \varepsilon f(\mathbf{p}_4)) f(\mathbf{p}_2) - \varepsilon f(\mathbf{p}_3) f(\mathbf{p}_4)(1 + \varepsilon f(\mathbf{p}_2))], \qquad (32)$$

with

$$f(\mathbf{p}_i) \equiv [\exp(\beta\{E(\mathbf{p}_i) - \mu\}) - \varepsilon]^{-1}.$$

This rate depends on the matrix element squared for scattering, times the density of scatterers, modified by statistical effects. For fermions with $|p| \sim p_F$, $\omega \sim E(p_F)/\hbar$, $\Gamma' \sim (kT/\varepsilon_F)^2$. As a result $\tau \sim (\varepsilon_F/kT)^2$. Since $\Gamma' \to 0$ for $\hbar\omega \to E(p_F)$ and $(|p| \sim p_F)$ it follows that $n(p)$, defined by (22) has a discontinuity.[75]

More precisely and generally we may state that since G^{-1} has no poles for complex z

$$\varrho(\mathbf{p}\omega) = \frac{\varGamma'(\mathbf{p}\omega)}{\left(\omega - \dfrac{E_{\text{H.F.}}(\mathbf{p})}{\hbar} + P\displaystyle\int \dfrac{d\omega'}{2\pi} \dfrac{\varGamma'(\mathbf{p}\omega')}{\omega' - \omega}\right)^2 + \left(\dfrac{1}{2}\varGamma'(\mathbf{p}\omega)\right)^2} . \quad (33)$$

If there is a unique significant solution $E(\mathbf{p})$ to the equations

$$E(\mathbf{p}) = E_{\text{H.F.}}(\mathbf{p}) - P\int \frac{d\omega'}{2\pi} \frac{\hbar\varGamma'(\mathbf{p}\omega')}{\omega' - E(\mathbf{p})/\hbar} \quad (34)$$

this solution, $E(\mathbf{p})$ is called *the* energy of the "quasi-particle." In the neighborhood of this energy we may write as in (A43)

$$\varrho(\mathbf{p}\omega) = \frac{\varGamma''(\mathbf{p})\, Z(\mathbf{p})}{(\omega - E(\mathbf{p})/\hbar)^2 + (\frac{1}{2}\varGamma'(\mathbf{p}))^2} + r(\mathbf{p}\omega) \quad (35)$$

where $r(\mathbf{p}\omega)$ is a smooth function $\varGamma''(\mathbf{p}) = \varGamma''(\mathbf{p}E(\mathbf{p}))\, Z(\mathbf{p})$ is smooth and slowly varying, and

$$Z(\mathbf{p}) = \left[1 + \frac{\partial}{\partial E(\mathbf{p})} P\int \frac{d\omega'}{\pi} \frac{\hbar\varGamma''(\mathbf{p}\omega')}{\omega' - (E(\mathbf{p})/\hbar)}\right]^{-1} . \quad (36)$$

If $\varGamma''$ and hence $\varGamma'$ approaches zero at $T = 0$, then there will be a discontinuity at the fermi surface of strength Z. (Fig. 30).

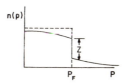

Fig. 30. The momentum distribution of a normal fermi system at zero temperature to all orders of perturbation theory. The dotted line shows the momentum distribution with no interaction.

The analogous classical development for the operator

$$f(\mathbf{r}\mathbf{p}t) = \sum_{\alpha} \delta(\mathbf{r} - \mathbf{r}^\alpha(t))\, \delta(\mathbf{p} - \mathbf{p}^\alpha(t)) \quad (37)$$

and the correlation function

$$\langle f(\mathbf{r}\mathbf{p}t) f(\mathbf{r}'\mathbf{p}'t')\rangle_{\text{eq.}}$$

and its "mass operator" (which describes the diffusion and oscillation of deviations of the one particle distribution in one particle phase space,) bear corresponding study but we cannot enter into it here. It is of course but one more illustration of the recurring discoveries made by different physicists with different backgrounds, working in different fields. The language of field theory, and of electrical engineering, in quantum systems and classical systems, for nuclear reactions, magnetic resonance, and scattering, even in lasers and superfluids, is really one language and it is that language which these lectures have attempted to codify.

References

Supported in part by a grant from the National Science Foundation.
[1] We use rationalized Gaussian units.
[2] The reader should not infer that the Nyquist theorem is really necessary or relevant in proving that the moments exist.
[3] This kind of phenomenological ansatz which corresponds to a force relaxing exponentially in time, has been used frequently over many years.
[4] Whether other peaks in $\chi''(\omega)$ should be called resonances is entirely a matter of taste and semantics. There is no reason to expect them to be Lorentzian.
[5] The reader should calculate $\chi''(\omega)$ for two coupled oscillators, determining the significance of λ and ω_0 in this case.
[6] A. Mooradian and G. B. Wright, *Phys. Rev. Letters* **16**, 999 (1966).
[7] A. Rahman, *Phys. Rev.* **136**, A405 (1964).
[8] L. Verlet, (to be published). We would like to thank Dr. Verlet for communicating his results prior to publication.
[9] The interpolation scheme and the figures will be published by S. Yip and P. C. Martin. The curve for $\gamma'(\omega)$ has been independently obtained by L. Verlet.
[10] B. J. Berne, J. P. Boon, and S. A. Rice, *J. Chem. Phys.* **45**, 1086 (1966).
[11] A discussion of localized modes and references to recent experiments can be found in the review article of A. Maradudin, *Solid State Physics* **18**, 273–420 (1966) and **19**, 1–134 (1966).
[12] Much of the material in this chapter (all save Sec. 6) was first presented from this general viewpoint by R. Kubo in a variety of articles. See for example Proc. Scottish Summer School 1961. Parts of it are also related to the work of L. van Hove, *Phys. Rev.* **95**, 249 (1954).
[13] L. P. Kadanoff and P. C. Martin, *Annals of Physics* (N.Y.) **24**, 419 (1963).
[14] This dispersion relation and the significance of the term in ω_0^2 is discussed at length by P. C. Martin, *Statistical Mechanics of Equilibrium and non-Equilibrium* edited by J. Meixner, North Holland, Amsterdam, (1965) p. 100. The quantity Γ is sometimes called the memory function. Clearly, the same analysis that leads from χ to Γ could also be applied to replace Γ by another function, etc. In this manner we would generate a continued fraction representation. Such representations have been advocated from another point of view in the same volume by H. Mori.
[15] B. R. A. Nijboer and A. Rahman, *Physica* **32**, 415 (1966).
[16] J. G. Kirkwood, *J. Chem. Phys.* **14**, 180 (1946).
[17] The first part of the section follows Ref. 13.
[18] The most recent data can be found in A. Anderson, *et al.*, *Phys. Rev. Letters* **17**, 367 (1966), where further references are given.
[19] Several experimental studies of this slowing down can be found in the *Proceedings of the Conference on Phenomena in the Neighborhood of Critical Points*, Washington, D.C. (1965), ed. by M. S. Green and J. V. Sengers NBS, *Misc. Publication* No. 273.
[20] This example is discussed in Ref. 14.
[21] These interpolation formula are extensively discussed and employed by K. F. Herzfeld and T. A. Litovitz, *Absorption and Dispersion of Ultrasonic Waves*, Academic Press, N.Y. (1959). They may also be found in L. D. Landau and E. M. Lifshitz, *Fluid Mechanics*, Addison-Wesley, Reading, Mass. (1959).

[22] The longitudinal formula analogous to (47) was obtained classically by P.G. de Gennes, *Physica* **25**, 825 (1959), and quantum mechanically by R. D. Puff, *Phys. Rev.* **137**, A406 (1965). The transverse formula is obtained as a corollary. It has been studied in classical systems (with $k = 0$) by R. Zwanzig and R. D. Mountain, *J. Chem. Phys.* **43**, 4464 (1965).

[23] To be published, Forster, Yip, and Martin.

[24] To be published, Forster, Yip and Martin.

[25] There is a vast literature dealing with the attenuation of shear waves in anharmonic crystals. The formula to which we refer was first derived by L. D. Landau and G. Rumer, Physik Z. Sowjetunion **11**, 18 (1937). A discussion in the terms we have introduced and the other references can be found in P. C. Kwok and P. C. Martin, *Solid State Communications* **3**, 181 (1965).

[26] Ref. 13.

[27] Such symmetries, which eliminate the coupling between quantities of different vectorial character, are what enable us to discuss the transverse momentum separately from the other degrees of freedom in a fluid. Additional terms vanish for special reasons (e.g. the fact that the particle currents in a fluid are linearly related to the momentum). Finally the matrix of coefficients has symmetry properties which further reduce the number of terms.

[28] The corresponding formulas for superfluids are derived in P. C. Hohenberg and P. C. Martin, *Annals of Physics* (N.Y.) **34**, 291 (1965).

[29] L. Landau and G. Placzek, *Phys. Z. Sowjetunion* **5**, 172 (1934).

[30] See, for example, N. Ford and G. B. Benedek, *Phys. Rev. Letters* **15**, 649 (1965).

[31] S. S. Alpert, Y. Yeh and E. Lipworth, *Phys. Rev. Letters* **14**, 486 (1965).

[32] Theoretical studies of this region have been carried out by S. Yip and several co-workers (M. Nelkin, S. Ranganathan, and J. van Leeuwen). Figs. 19 and 20 are taken from Nelkin, van Leeuwen, and Yip, *Inelastic Scattering of Neutrons*, Vol. II, *International Atomic Energy Agency*, Vienna, 1965, p. 35. The theoretical studies on Fig. 21, and more recent ones which use a hard core interaction and give better agreement with experiment have been provided by S. Yip and are not yet published.

[33] The experimental studies are reported by T. J. Greytak and G. B. Benedek, *Phys. Rev. Letters* **17**, 199 (1966). We would like to thank Dr. Greytak for providing us with the more complete data in his thesis.

[34] This topic is also the subject of an extensive literature. In helium the most extensive calculations have been performed by I. M. Khalatnikov and D. M. Chernikova, *J. Exptl. Phys.* U.S.S.R. **50**, 411 (1966), *Soviet Physics* **23**, 274 (1966), Experiments in the phonon region have been carried out by B. M. Abraham, *et al.* *Phys. Rev. Letters* **16**, 1039 (1966). The topic has also been discussed by C. J. Pethick and D. ter Haar, *Physica* **32**, 1905 (1966), and by Kwok and Martin, Ref. 25 where further references are given for solids as well.

[35] J. C. Ward and J. Wilks, *Phil. Mag.* **43**, 48 (1952). The correlation function approach has been discussed by P. C. Kwok and P. C. Martin, *Phys. Rev.* **142**, 495 (1966), L. Sham, *Phys. Rev.* **156**, 494 (1967), and W. Götze and K. Michel, *Phys. Rev.* **156**, 963 (1967). See also R. A. Guyer and J. Krumhansl, *Phys. Rev.* **148**, 766, 778, 789 (1966).

[36] C. C. Ackerman, *et al.*, *Phys. Rev. Letters* **16**, 789 (1966).

[37] M. A. Woolf, P. M. Platzman, and M. G. Cohen, *Phys. Rev. Letters* **17**, 294 (1966).

[38] The familiar dispersion curve for the excitations in liquid helium may be found in any discussion of helium. See, for example, L. D. Landau, and E. M. Lifshitz, *Statistical Physics*, Addison-Wesley, Reading, Mass. (1958), p. 198.

[39] Results on a variety of simple liquids have been reported by S. H. Chen *et al.*, *Physics Letters* **19**, 269 (1965).

[40] Recent experiments on liquid lead seem to indicate a similar phonon-like behavior. See P. D. Randolph and K. S. Singwi, *Phys. Rev.* **152**, 99 (1966).

[41] Ref. 13 shows the equivalent

$$\lim_{\omega \to 0} \lim_{k \to 0} \omega \chi_L''(k\omega)/k^2 = \zeta + \frac{4}{3}\eta$$

Note that as $k \to 0$, $\omega \chi''(k\omega)/k^2 \to \left(\zeta + \frac{4}{3}\eta\right) + \pi \left(\frac{dp}{dn}\right)_s n\delta(\omega)$

$$mnD_L'(k\omega) \to \left(\zeta + \frac{4}{3}\eta\right) + \pi mn \left(\frac{dp}{dn}\right)_T \left(\frac{c_p}{c_v} - 1\right)\delta(\omega)$$

The terms in $\delta(\omega)$ do not contribute to the ω-limits although the time independent terms they correspond to make the time integral of the correlation function diverge.

[42] The form (15) guarantees that

$$\int \frac{d\omega}{\pi} \frac{\chi_L''(k\omega)}{\omega mn} = \int \frac{d\omega}{\pi} \frac{m}{n} \frac{\omega}{k^2} \chi_{nn}''(k\omega) = 1.$$

Together with the compressibility sum rule and the assumption that

$$\chi_L''(k\omega)/\omega = mn\,\pi\delta(\omega^2 - \omega_{LO}^2(k))\omega_{LO}(k)$$
$$= mn\,\pi\delta(\omega^2 - k^2 c_{LO}^2(k))\,c_{LO}(k)\,k$$

we find that, if there is a resonance at $\omega_{LO}(k)$, as $k \to 0$, $\omega_{LO}^2 = c_{LO}^2 k^2 \to (dp/dmn)k^2$. The relation $c_{LO}^2(k) = (n/m)\chi_{nn}^{-1}(k0)$ agrees with this expression as $k \to 0$. When $k \ne 0$, the resonance assumption and the fluctuation dissipation theorem imply that $S_{nn}(k\omega) \cong 2\hbar\eta(\omega)\,\chi_{nn}''(k\omega)$ for $\beta\hbar\omega \gg 1$ so that

$$\tilde{S}_{nn}(k0) = \int_{-\infty}^{\infty} \frac{d\omega}{2\pi} S_{nn}(k\omega) = \hbar \int_0^{\infty} \frac{d\omega}{\pi} \chi_{nn}''(k\omega) = \frac{\hbar}{2}\chi_{nn}''(k0)\,\omega_{LO}(k) = \frac{\hbar nk^2}{2m\omega_{LO}(k)};$$

they also imply $S_{nn}(k\omega) \cong (2/\beta\omega)\,\chi_{nn}''(k\omega)$ for $\beta\hbar\omega \ll 1$ so that

$$\tilde{S}_{nn}(k0) = \frac{\chi_{nn}(k0)}{\beta} = \frac{nk^2}{\beta m\omega_{LO}^2(k)}.$$

Both forms indicate that $\omega_{LO}(k)$ will have a minimum for that value of k for which $\tilde{S}_{nn}(k0)$ has a maximum. This qualitative feature appears to occur both in helium and the noble gas fluids. The first of the two formulas which make this prediction is called the Feynman-Bijl formula.

Note that if we were to truly take the resonance assumption seriously for all k, and if we were to know the kinetic energy these formula and (17) would give a closed integral equation for $\tilde{S}_{nn}(k0)$ since its fourier transform determines the function $g^{(2)}(r)$ occurring in (17). Such equations have been studied by Puff (Ref. 22) but they are not very useful because large k (for which the ansatz is poor) play an important role in the integral equation.

[43] P. C. Hohenberg and P. C. Martin, *Ann. Phys.* (N. Y.) Ref. 34.

[44] See N. Ford and G. B. Benedek, Ref. 19.

[45] The graph of kinematics is taken from the review article by P. K. Iyengar in *Thermal Neutron Scattering*, ed. by P. A. Egelstaff, Academic Press, New York (1965). A more detailed discussion of experimental techniques may be found there.

[46] This discussion may be found in a large number of reviews of neutron scattering. The original source is L. van Hove, Ref. 12.

[47] The results displayed are taken from B. N. Brockhouse and N. K. Pope, *Phys. Rev. Letters* **3**, 259 (1959) and B. A. Dasannacharya and K. R. Rao, *Phys. Rev.* **137** A, 117 (1965).

[48] A. Einstein, Annal. Physik. **38**, 1275 (1910). (The factor of $\sqrt{\bar{\varepsilon}}$ in the specific form quoted takes account of the fact that the light moves more slowly in the medium.)

[49] Among the more recent studies of critical scattering in ferromagnets are those of B. Jacrot, et al., *Inelastic Scattering of Neutrons in Solids and Liquids*, 317, IAEA, Vienna (1963), and L. Passel, *et al. J. Applied Physics* **35**, 933 (1964), *Phys. Rev.* **139A**, 1866 (1965); in antiferromagnets, B. H. Torrie, *Proc. Phys. Soc.* **89**, 77 (1966) and M. J. Cooper and R. Nathans, *J. Applied Phys.* **37**, 1041 (1966).

[50] The behavior of the diffusion constant in a ferromagnet near T has been studied by B. Jacrot *et al.*, and L. Passel, *et al.*, Ref. 49. Recent studies have been performed on antiferromagnets by Nathans, Menzinger, and Pickart (to be published).

[51] P. A. Egelstaff, *Thermal Neutron Scattering*, Academic Press, New York (1965).

[52] Much of the material in this chapter can be found in P. C. Martin, *Phys. Rev.* **161**, 143 (1967).

[53] E. M. Purcell, Berkeley Physics Course Vol. 2: *Electricity and Magnetism*. McGraw Hill, New York (1963).

[54] L. P. Kadanoff and P. C. Martin, *Phys. Rev.* **124**, 677 (1961).

[55] This distinction between total field and external field is the source of the various difficulties that arise in discussing charged systems.

[56] The occurrence of Friedel oscillations in the screened exchange results from the singularity at $k = 2k_F$.

[57] The difficulty with causality arguments and the importance of recognizing that the total field is not necessarily arbitrary is discussed in Ref. 52.

[58] This transverse sum rule has been employed in analyzing infrared absorption in superconductors by M. Tinkham and R. Ferrell, *Phys. Rev. Letters* **2**, 331 (1959).

[59] While magnetic screening occurs in superconductors it is frequently too rapid for the London equations to apply. The k-dependence of the quantity $\mu^{-1}(k)$ becomes important at a coherence length ξ, and when the London penetration depth k_L^{-1} is less than ξ the penetration depth actually lies between k_L^{-1} and ξ. This is known as the Pippard limit.

[60] The superfluid sum rules are discussed in Refs. 13, 14, and more extensively in Ref. 28. The difficulty in making the moment of inertia less than the rigid moment without some phenomenon like superconductivity was a dilemma in the nuclear theory of aspherical nuclei for some time.

[61] These expressions are ordinarily written, like (C14) or (C36), as time integrals of the current correlations.

[62] These integrals correspond to the sum rules (D10) and (D17).

[63] Actually there is an additional restriction on the validity of the time dependent results. They are only correct for times short compared to τ where τ is the long time determined by $(nv\sigma)^{-1}$ where σ is the screened Coulomb scattering cross section. The low frequencies associated with these long times are unimportant in frequency integrals but important for discussing transport properties of the plasma.

[64] A long but very instructive exercise is to repeat the steps in this section for a fermi system in its ground state. From its $\chi_{nn}^{0\prime\prime}(k\omega)$ the screening, plasma dispersion relations, and the dynamic properties may be obtained.

[65] The classical energy loss calculation with these cutoffs is due to N. Bohr, *Phil. Mag.* **24**, 10 (1913), **30**, 581 (1915) and may be found in a variety of elementary texts. Note that the mass, M, of the slowed particle does not appear.

[66] A second quantum cut off is provided by the exponential factor. It restricts k to values less than or approximately equal to $p/\hbar$. This cut off is also present in (51) but absent from (52).

[67] Of course, in our free electron model taken literally, the index of refraction is never greater than unity and there is no Čerenkov radiation. The same formulation is applicable in systems with bound charge whose index of refraction is larger than one for an interesting range of frequencies.

[68] The same technique has been employed to calculate the correlation energy (of order $\log r_s$, where $r_s \sim me^2/\hbar k_F$) in a charged fermi gas by K. Sawada, et al., *Phys. Rev.* **108**, 507 (1957).

[69] P. Debye and E. Hückel, *Phys. Z.* **24**, 185 (1923).

[70] For fermions, the source term in the hamiltonian, which represents another fermi field should be taken to anticommute with the fermi field ψ. These anticommuting sources are familiar in field theory and were introduced by J. Schwinger, *Proc. Natl. Acad. Sci. U.S.* **37**, 952 (1951).

[71] The fluctuation dissipation theorem can also be formally expressed by the statement $G_>(rt; r't') = \varepsilon G_<(rt + i\beta\hbar; r't')$ with $H \to H - \mu N$. This expression in terms of a periodic property of the correlation functions plays a central role in the techniques developed by P. C. Martin and J. Schwinger, *Phys. Rev.* **115**, 1342 (1959) and Abrikosov, Gor'kov and Dzyaloshinski, *Field Theory Methods in Quantum Statistical Mechanics*, Prentice Hall, Englwood Cliffs, New Jersey (1963).

[72] The parameter n_0 should not be confused with the parameter n_s introduced earlier. The latter is measurable and greater than the former except at $T = 0$ in a perfect bose gas where both equal unity, and at and above the transition temperature where they both vanish.

[73] Although the formula is written for a translationally invariant system, obviously the Hartree-Fock approximation, and its finite temperature generalization, are not restricted to such systems.

[74] Although the formula is plausible a derivation requires some examination of perturbation theory. The derivation may be found, for example, in L. P. Kadanoff and G. Baym, *Quantum Statistical Mechanics*, Benjamin, New York (1962), p. 36.

[75] That the width of ϱ approaches zero at the fermi energy and consequently $n(p)$ is discontinuous may be verified in perturbation theory. The discontinuity is not present in a superconducting system, and one may suppose that probably all systems have some such transition at sufficiently low temperatures. Nevertheless, the fact that such widths are small at low temperatures on a practical scale, is demonstrated by the sharp fermi surfaces at low temperatures in de Haas-van Alphen studies.